An Introduction to Polysaccharide Biotechnology

An Introduction to Polysaccharide Biotechnology

By

Stephen E. Harding, Michael P. Tombs and Gary G. Adams
University of Nottingham, UK

and

**Berit Smestad Paulsen, Kari Tvete Inngjerdingen
and Hilde Barsett**
University of Oslo, Norway

2nd Edition

CRC Press
Taylor & Francis Group
Boca Raton London New York

CRC Press is an imprint of the
Taylor & Francis Group, an **informa** business

CRC Press
Taylor & Francis Group
6000 Broken Sound Parkway NW, Suite 300
Boca Raton, FL 33487-2742

First issued in paperback 2017

ISBN 13: 978-0-815-38715-2 (pbk)
ISBN 13: 978-1-4822-4697-1 (hbk)

Library of Congress Cataloging-in-Publication Data

Names: Harding, S. E. (Stephen E.)
Title: An Introduction to Polysaccharide Biotechnology.
Description: Second edition / Stephen E. Harding [and five others]. | Boca
Raton : CRC Press, [2017] | Includes bibliographical references and index.
Identifiers: LCCN 2016040404 | ISBN 9781482246971 (hbk : alk. paper) | ISBN
9781482248326 (ebk : alk. paper)
Subjects: LCSH: Polysaccharides--Biotechnology. |
Polysaccharides--Metabolism. | Glucosidases--Biotechnology.
Classification: LCC TP248.65.P64 T66 2017 | DDC 660.6/3--dc23
LC record available at https://lccn.loc.gov/2016040404

Visit the Taylor & Francis Web site at
http://www.taylorandfrancis.com

and the CRC Press Web site at
http://www.crcpress.com

Contents

Preface to First Edition

Many universities have recently introduced courses with a major content of biotechnology. In addition, many of the older courses in biological sciences have found it desirable to incorporate material which could properly be described as biotechnology. This book is written for students of such subject areas, but we hope will also be found useful by others who want a compact but reasonably comprehensive account of the rapidly growing biotechnology of polysaccharides.

Because it is undergoing such rapid growth, there will almost certainly be many examples of polysaccharide biotechnology in actual application when this book appears which had not reached that stage when it was written. For this reason, we have allowed ourselves to include some examples which look as if they are likely to find application in the near future, despite our endeavour to illustrate the technology by examples that are actually in use, rather than possible processes that might one day find application, but all too often really illustrate unfulfilled promise.

We have taken a broad view of the scope of biotechnology. In its early days, one might have got the impression that it consisted of large-scale fermentations for the production of a variety of small molecules. It is now far beyond that and could be broadly considered as any applied field for which more recent developments in molecular biology provide the underlying basis. Generally if there is a doubt we have included it.

A useful working definition of biotechnology is that it is what people calling themselves biotechnologists do. Our material would pass that test. There is a selection of references in the text, and they are chosen to be representative rather than particularly notable. More important we give some general references – mostly reviews. With the ready availability of databases, any student can find a more comprehensive list of references to the primary literature than we could ever include, given a few leads, and we have chosen not to clutter up the text by the inclusion of dozens of references included on a more or less ritual basis which will in practice never be used. Much better to give a way in to the necessary key words, which will probably survive an out of date reference list.

A major problem in biotechnology at present is that the useful properties of macromolecules, especially polysaccharides, can rarely be specified in terms of the molecular properties. On the other hand, molecular biology and the results of genetic manipulation are almost always described in molecular terms. For example, it is possible by suitable gene modification to alter the degree of branching of many polysaccharides while precisely what effect this will have on a product on the supermarket shelf is not predictable or even describable in such terms.

We have the slightly absurd situation, at first glance, that the results of complex DNA manipulation, and extremely detailed chemistry, may have to be assessed by a taste panel. Taste panels are not very good at this. There would be fewer new product failures if they were better, but there is all too often nothing else available. Much current effort is going to fill the gap for the food industry, and the situation is less bad in, for example, pharmaceuticals where the molecular basis of action is better understood.

For this reason, we have included a substantial amount on the characterization of polysaccharides, because this will offer the only way of assessing the suitability of novel materials for their potential use. This is largely based on their properties in solution because this is where most of them will find application. The contrast with heterologous expression of enzymes is obvious, since in that case the activity of the enzyme is central to the novel application, while this easy test is completely lacking in the case of polysaccharides. They must be assessed on a different basis.

Each chapter includes a brief account of the basic chemistry, including size and shape, of the polysaccharide or group of polysaccharides and then sets the industrial scene with an account of uses and in some cases existing biotechnology-based processes. We then consider the biosynthetic pathways and opportunities for the modification of the molecular structures and the way in which this has been used, or in a few cases might be used to enhance the applications. Most polysaccharides derive from plants – the overwhelming bulk of those that have found useful applications certainly do – and so we have included modification to the source crops whether or not this has been done by conventional plant breeding.

Finally, we would like to thank Gudmund Skjåk Braek and Bjørn-Torger Stokke for their comments and suggestions. We must also apologize in advance for the omissions which are inevitable in such a fast-moving field with a diverse literature and worldwide interest. We have tried to include everything but there will certainly be something that we have missed.

Michael P. Tombs, Stephen E. Harding
Nottingham

Preface to Second Edition

It has been nearly 20 years since the publication of the first edition. These two decades have seen two parallel developments. First, a huge growth in the application of carbohydrate-based technologies and second, an expansion of biotechnology-related courses taught at universities. At the Universities of Nottingham and Oslo for example we have seen in particular the growth of courses on biotechnology at not only undergraduate level but in particular at postgraduate Masters level as well, and also courses taken by PhD/DPhil students as part of their training. In this new edition, we have kept the successful format of the first edition but updated the content to reflect the growth of the subject, introducing students to new developments and applications in, for example, marker-assisted selection and other genetic technologies. We also consider the growth in (and our understanding of) the application of polysaccharide bioactivity, and a whole new chapter is devoted to this important area. This new edition also introduces students to the fascinating and hugely important area of polysaccharide and glycoconjugate vaccines against serious disease such as meningitis. Whilst keeping the book at an introductory level, key follow-up references to topics are given. *An Introduction to Polysaccharide Biotechnology* should be of interest not only to students but also to other new researchers and even experienced scientists. The student should be aware that there have recently been some changes in the nomenclature of plants – the sources of many polysaccharides used in biotechnology. Throughout the book, we have used the plant names corresponding to the time when the papers referred to were published. Finally, we would like to thank Professor Graham Seymour for his comments and suggestions on this new edition.

Stephen E. Harding, Michael P. Tombs, Gary G. Adams,
Berit Smestad Paulsen, Kari Tvete Inngjerdingen and Hilde Barsett
Nottingham and Oslo

Note for Second Edition

Following student feedback to the previous edition, in this current edition each chapter contains a Further Reading section as a follow up source for students, in addition to the Specific Reference section.

Polysaccharides and Their Potential for Biotechnology

1.1 AN INTRODUCTION TO POLYSACCHARIDES

Over the last two decades, there has been a rapidly expanding interest in polysaccharides and glycosubstances from both a *fundamental* viewpoint – with regards to the crucial role they play in molecular recognition phenomena – and also from an *applications* viewpoint, in terms of their considerable biotechnological usefulness. This latter feature derives from the enormous diversity of structural and functional properties even though they are built up from very similar building blocks: the pyranose or furanose carbohydrate ring structure. With the polysaccharides, one can have extended rod-shape molecules such as xanthans and schizophyllan, linear random coil-type structures such as dextrans and pullulans, branched structures such as glycogen and amylopectin, polyanions such as alginates, pectins, carrageenans, xanthans and hyaluronic acid, neutral structures such as guar, pullulan and dextran, and finally also polycations such as aminocelluloses and chitosans. Virtually all of these molecules are non-toxic and harvestable at low cost in large quantities.

It is usual when describing polysaccharides to begin by pointing out that they are quite different from that other major class of biological polymers, the proteins. Proteins are built up from at least 20 different amino acid residues, and most proteins contain all of them. Polysaccharides are sometimes homopolymers, built from a single kind of pentose or hexose residue, and even when more than one is present this rarely exceeds three or four. Proteins can form tightly folded structures, while polysaccharides never produce anything similar. Instead many approximate much more closely the idealized random-coil configurations originally defined for simple synthetic homopolymers. Most proteins do not, except in carefully chosen solvents which eliminate most of the side chain–side chain interactions on which the native structure depends. Some proteins – of which the most important example is gelatin – do not possess detailed folded structures and behave more like random or extended coils in solution. Indeed gelatin has been described as an 'honorary polysaccharide' because its hydrodynamic behaviour is much more like that of many polysaccharides than that of a typical globular protein. Polysaccharides can, however, form defined structures. In particular, helices are often found, especially in the more insoluble species. Both proteins and carbohydrates can carry charged groups of either sign, though only the synthetic polyamino acids like polyleucine could be compared with the neutral polyglucoses like starch. They share the common property of a very low solubility in water. By far the most important difference, however, especially for the hydrodynamic properties, is the frequent occurrence of branched chains in polysaccharides. These are virtually unknown in proteins, and even the nearest equivalent, interchain disulphide links, which could be regarded as a form of branching, are fairly uncommon. Another difference is the way in which pentoses and hexoses can link to form chains in different ways, rather than being confined to the standard peptide link or the slightly different one (involving proline) found in proteins.

Functions too are different. Polysaccharides are either structural components, most notably in plant cell walls, or storage components, particularly in seeds. It is possible that some have a minor and probably secondary role as wound healing agents, or even anti-microbial activity, since they appear typically when plants are damaged, but even so they seem mostly to be involved as stores of carbohydrate. In some specialized situations, such as the anchors which fix algae to rocks, they may function as adhesives. They may become involved in lectin interactions on the surface of cells, though there are no clear examples of this *in vivo* except possibly for some bacterial polysaccharides. There is never any suggestion that the sort of catalytic activity associated with enzymes can be brought about by polysaccharides. Even in this respect, however, it is possible to use cyclic dextrans as catalysts in some reactions, though it is stretching the case too far to see this as an indication of a catalytic role *in vivo*. A new class of synthetic polysaccharides – the aminocelluloses – is proving extremely interesting because of the protein-like properties (reversible oligomerization reactions and globular-like properties) (Heinze et al., 2011; Nikolajski et al., 2014) and with the potential to form transparent films. These films can be linked to functional groups such as in enzymes allowing for example bifunctional enzyme coupled reactions (Jung and Berlin, 2005).

1.1.1 GLYCOSUBSTANCES AND GLYCANS

Glycoproteins and mucosubstances contain both polysaccharide and protein chains covalently linked together. It is important to emphasize the covalent nature of the structure, since over the last 30 years many materials which were thought to be mucosubstances have proved simply to be mixtures. Hyaluronic acid, for example, when isolated by one method from synovial fluid contains a fairly constant level of associated protein, which may actually be part of the functional entity *in vivo*. A modified method of isolation leads to the removal of the protein component and hyaluronic acid would not now be regarded as a mucosubstance. In the past, a distinction was made between glycoproteins, which contain relatively little carbohydrate – an example would be ovalbumin with only one or two hexose residues per molecule – and mucosubstances, which were viewed as containing much more carbohydrate, typically 30% or more. The archetypes are orosomucoid (~50% glycosylated) found in blood or the elastins (~50%) found in plant cell walls, or the 'mucins', the highly glycosylated (~80%) glycoproteins, which give mucus its characteristic high viscosity

E. Fischer (Nobel Prize, 1902)   W. N. Haworth (Nobel Prize, 1937)

and gel forming characteristics, characteristics vital for its bioprotective role. This distinction, however, has now largely fallen into disuse – and was in any case never sharply defined – so all these molecules tend now to be called glycoproteins, glycoconjugates (a more general term allowing for non-protein as well as protein [covalently] linked entities to the carbohydrate) or simply *glycosubstances*. The preferred modern term for describing polysaccharides and glycoconjugates together is *glycans*. Total glycopolymer and polysaccharide nomenclature has, however, still not entirely settled down and the student should be prepared to find all these terminologies in use, especially in the more anarchic world of commercial biotechnology. In this book, we are concerned mostly with polysaccharides and other high molecular weight glycans containing substantial amounts of polysaccharide chain.

1.1.2 STRUCTURES, NOMENCLATURE AND ABBREVIATIONS

Like proteins, polysaccharides have a structural hierarchy: primary structure is the monosaccharide composition; secondary structure is the local arrangement of groups of oligosaccharide units in a polysaccharide chain; tertiary structure is the conformation of a polysaccharide chain; and quaternary structure is the structure of any interchain complex or complex with another macromolecule or ligand.

Monosaccharide building blocks. These are either hexoses (six carbon atoms) or pentoses (five carbon atoms) and have the following features:

1. By analogy with amino acids, they can exist as two enantiomorph forms ('D' and 'L') depending on which direction their parent or unsubstituted substance rotates plane-polarized light (respectively, clockwise or anticlockwise). Whereas with amino acids the L form is the normal naturally occurring form, with monosaccharides it is usually the D form although there are more exceptions. Further notation is often used, '+' and '−', to indicate the actual rotation of the substance, whether it be the parent monosaccharide or a substituted variant thereof.
2. They are either aldoses (R–CHO) or ketoses (R–CO–R′) as represented by the 'Fischer' (after Emil Fischer who won the Nobel Prize for Chemistry for this work in 1902) representations or 'projection' of the structure (Figure 1.1).

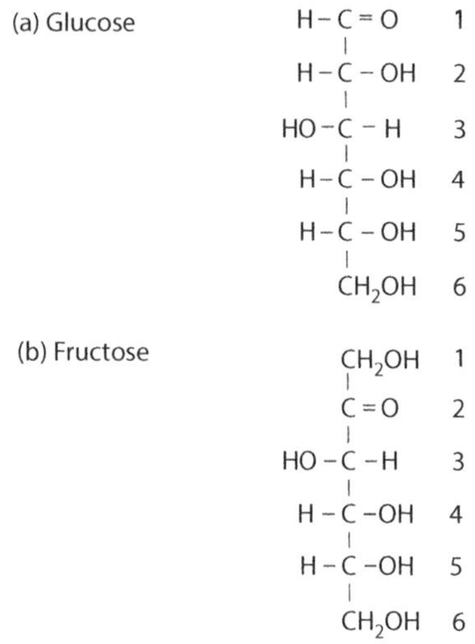

(a) Glucose

$$\begin{array}{ll}
H-C=O & 1 \\
H-C-OH & 2 \\
HO-C-H & 3 \\
H-C-OH & 4 \\
H-C-OH & 5 \\
CH_2OH & 6
\end{array}$$

(b) Fructose

$$\begin{array}{ll}
CH_2OH & 1 \\
C=O & 2 \\
HO-C-H & 3 \\
H-C-OH & 4 \\
H-C-OH & 5 \\
CH_2OH & 6
\end{array}$$

FIGURE 1.1 Fischer structures for an aldose (D-glucose) and a ketose (D-fructose). These structures are isomers with molecular weight $M = 180$ Da.

3. They cyclize via a hemiacetal linkage to give a ring structure (Figure 1.2) either in the pyranose 'p' form, a six-membered ring, including oxygen, or in the furanose 'f' form, a five-membered ring, including oxygen. W. N. Haworth who, like Fischer earlier, won the Nobel Prize for Chemistry in 1937, first presented this ring representation. Groups on the left-hand side (LHS) and right-hand side (RHS) of the backbone – C – atoms in the Fischer projection appear, respectively, above and below the 'plane' of the ring structure.

4. The orientation of the OH group on Cl (pyranose) or C2 (furanose) can take one of two forms represented by the notation 'α' or 'β'. For example in the representation of Figure 1.2, for α-D-glucose the OH group on Cl is below the ring, whereas for β-D-glucose it is above as indicated by the arrow.

5. The Haworth structure of Figure 1.2 does not tell the complete story about the stereochemistry of the monosaccharide: the ring structure cannot be completely planar because of the tetrahedral nature of most of the bonds involving the C atoms. Students who have done organic chemistry may be aware of the 'chair' and 'boat' steric configurations of cyclohexane, with the chair being by far the more stable. By analogy, the pyranose ring in saccharides can adopt one of two stable configurations, both in the chair form, which one depending on whether the saccharide is a D- or L-sugar: these are called, respectively, the $^4C_1$ and $^1C_4$ forms (Figure 1.3): in both cases, the sterically favourable situation of locating the bulky CH_2OH (or derivative thereof) residue equatorially and away from other substituents which are generally situated axially is achieved.

6. The various names assigned, 'glucose', 'fructose', 'guluronic acid' and so on reflect both the relative orientation of the OH and H groups coming off the other ring C atoms and the presence

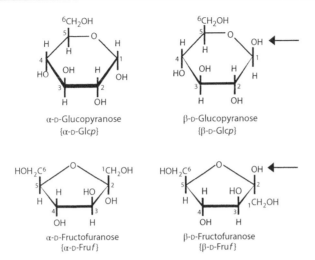

FIGURE 1.2 Haworth structures for pyranose and furanose rings of D-glucose.

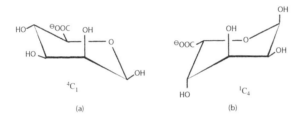

FIGURE 1.3 Preferred pyranose three-dimensional ring conformations, illustrated by residues found in alginate. (a) $^4C_1$ chair conformation for β-D-mannuronate, (b) $^1C_4$ chair conformation for α-L-guluronate.

and orientation of any other functional groups such as COO^-, NH_3^+, SO_3^-, CONH and so on. These names are most conveniently abbreviated to three to six letters: Glc for glucose, Xyl for xylose, Fru for fructose, GlcNAc for *N*-acetylglucosamine and so on (Table 1.1).

Di-, oligo- and polysaccharides. A disaccharide is formed from the formation of a glycosidic bond between two residues with a concomitant loss of a water residue. The most commonly known and most widely available is sucrose, one of very few pure substances bought directly over the counter. The molecular weight, *M*, of sucrose is 342 Da, and its formula in the now generally accepted saccharide notation is

$$α\text{-D-Glc}p\text{-}(1{\rightarrow}2)\text{-}β\text{-D-Fru}f$$

The Haworth structure for this is shown in Figure 1.4: as a warning to the reader some texts give incorrect representations. The student can find correct representations of this and other disaccharides in the excellent book by Candy (1980). In this notation, the (1→2) means the links are between the first C atom, with its OH in the α position, of

TABLE 1.1   Monosaccharide Nomenclature

Residue	Symbol	Residue	Symbol
Aldohexoses		*Amino sugars*	
Glucose	Glc	Sialic acid	
Galactose	Gal	(N-acetylneuraminic acid)	Neu5Ac
Mannose	Man	N-acetylglucosamine	GlcNac
		N-acetylgalactosamine	GalNac
Ketohexoses		*Uronic acids*	
Sorbose	Sor	Glucuronic acid	GlcA
Fructose	Fru	Galacturonic acid	GalA
Deoxyhexoses		Mannuronic acid	ManA
Rhamnose	Rha	Guluronic acid	GulA
Fucose	Fuc		
Aldopentoses			
Ribose	Rib		
Xylose	Xyl		
Arabinose	Ara		

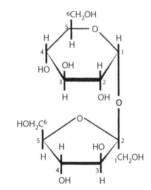

α-D-glucopyranosyl (1 → 2) β-D-fructofuranoside

{α-D-Glc$p$ (1 → 2) β-D-Fru$f$}

FIGURE 1.4   Haworth structure of sucrose.

D-glucose (pyranose form) and with the second C atom, with its OH in the β position, of D-fructose (furanose form). The arrow indicates the direction of the saccharide chain although the student should be warned that polysaccharide chemists for brevity often do not use the directional arrows → but use a hyphen - or comma , instead. An alternative way of writing this is [D-Glc$p$-(α1→β2)-D-Fru$f$].

In a similar way, and following the same notation scheme to represent them, further glycosidic bonds can be made to produce *oligosaccharides* consisting of typically between 3 and 10 monosaccharide residues and *polysaccharides* (consisting of >10 residues, or **degree of polymerization** [DP] >10). The student should note that in evaluating the molecular weight from the monosaccharide composition, care must be taken to allow for the loss of water residues. The sequences can be detected by sequential enzyme degradation and high-resolution nuclear magnetic resonance spectroscopy (Grasdalen, 1983). The primary sequence can be written as an extended form of the disaccharide notation above. However, unlike polypeptides which are copolymers of up to 23 different amino acid residues, polysaccharides rarely contain more than three different monosaccharide residues or linkages, and often contain just two or one (Figure 1.5): much simpler 'repeat' notation suffices. For example:

Linear:
  Amylose: $[\alpha\text{-D-Glc}p\text{-}(1{\rightarrow}4)]_n$ or $...{\rightarrow}4)\ \alpha\text{-D-Glc}p\text{-}(1{\rightarrow}...$
  Cellulose: $[\beta\text{-D-Glc}p\text{-}(1{\rightarrow}4)]_n$ or $...{\rightarrow}4)\ \beta\text{-D-Glc}p\text{-}(1{\rightarrow}...$

Alternating repeat:
  κ-carrageenan:
    $[\beta\text{-D-Gal}p\text{-4-sulphate }(1{\rightarrow}4)3,6\text{-anhydro-}\alpha\text{-D-Gal}p\ (1{\rightarrow}3)]_n$
    or $...{\rightarrow}3)\beta\text{-D-Gal}p\text{-4-sulphate }(1{\rightarrow}4)3,6\text{-anhydro-}\alpha\text{-D-Gal}p\text{-}(1 \rightarrow ...$

Interrupted repeat:
  Pectin:
    $...[\alpha\text{-D-Gal}p\text{A-}(1{\rightarrow}4)]_1 ... (1{\rightarrow}2)\text{-L-Rha}p\text{-}(1{\rightarrow}4)\text{-}[\alpha\text{-D-}$
    $\text{Gal}p\text{A-}(1{\rightarrow}4)]_n...$

Block copolymer:
  Alginate:
    $[\beta\text{-D-Man}p\text{A-}(1{\rightarrow}4)]_l[\alpha\text{-L-Gul}p\text{A-}(1{\rightarrow}4)]_m$
    $[\beta\text{-D-Man}p\text{A-}(1{\rightarrow}4)\alpha\text{-D-Gul}p\text{A-}(1{\rightarrow}4)]_n$

Branched copolymer:
Galactomannan: $...{\rightarrow}4)\ \beta\text{-D-Man}p\text{-}(1{\rightarrow}4)\ \beta\text{-D-Man}p\text{-}(1{\rightarrow}...$
$$6$$
$$\uparrow$$
$$1$$
$$\alpha\text{-D-Gal}p$$

Other structures include complex repeat types such as xanthan which contains at least three different monosaccharide residues (Figure 1.5).

Before leaving the subject of notation, it is worth pointing out that single letter notation is also very popular and useful, particularly when representing the sequence of residues in polysaccharide copolymers such as alginate and the galactomannans.

Polysaccharides are generally – but not exclusively – named according to their monosaccharide constitution (glucans for glucose, mannans for mannose etc.), prefixed by terms describing the linkage (e.g. beta glucans) or the presence of other residues (e.g. galactomannans), but there are no hard-and-fast rules. In terms of molecular weight, since the molecular weight of a monosaccharide unit is ~200 Da, the lower limit for a polysaccharide molecular weight is ~2000 Da (where a 'molecular weight' of 1 Da, as used by biochemists, is equivalent to a 'molar mass' of 1 g mol^{-1}, as used by physical chemists, where 1 mol is $6.022\,045 \times 10^{23}$ molecules). The situation is therefore analogous to that with polypeptides, where the molecular weight of the building block (amino acid) residue is of the same order (100–200

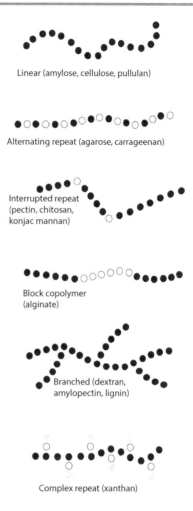

Linear (amylose, cellulose, pullulan)

Alternating repeat (agarose, carrageenan)

Interrupted repeat
(pectin, chitosan,
konjac mannan)

Block copolymer
(alginate)

Branched (dextran,
amylopectin, lignin)

Complex repeat (xanthan)

FIGURE 1.5 Polysaccharide primary structural types. (Adapted from Taylor, R.J. et al., *Carbohydrates*, Unilever, London, 1983. With permission.)

Da), 2 residues is a dipeptide, 3–10 residues is an oligopeptide (or simply 'peptide') and >10 residues (>~1500 Da) is a polypeptide.

Again, by analogy with a polypeptide, the 'secondary structure' or structure of oligosaccharide stretches within a polysaccharide is limited by the rotational degree of freedom of the glycosidic bond. As with the peptide bond, for glycosidic bonds between ring carbons (1→1, 1→2, 1→4 etc.) the rotational freedom is specified by two angles, ϕ and ψ (Figure 1.6). The rotational freedom is often severely limited by steric hindrance and also – for the heavily charged polysaccharides such as alginate, xanthan, chitosan and carrageenans – by electrostatic repulsion forces. For those polysaccharides involving linkage to a C6 atom, there is further rotational freedom specified by the angle ω: extensive 1→6 links in a polysaccharide tend to favour randomly coiled or 'disordered' conformations in solution – a good example is dextran.

In the solid state (fibres, films etc.), these angles are usually consistent for all or most glycosidic bonds in the polysaccharide chain: this gives rise

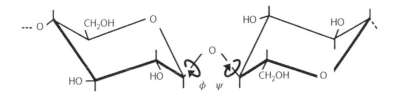

FIGURE 1.6   Rotational angles of the glycosidic bond.

to the so-called ordered state. This is also found under certain conditions in solutions of some polysaccharides (e.g. xanthan, carrageenan) and can promote the non-covalent self-association of polypeptide chains to produce double or even triple helices or (e.g. with pectins and alginates in the presence of calcium ions) 'egg-box dimers': all these conformational forms can be referred to loosely as 'tertiary structure'. Here, the parallel notation with proteins starts to depart since tertiary structure in proteins refers to the structure of a single polypeptide chain. Polysaccharide chemists usually use the term 'quaternary structure' to describe supermolecular assemblies such as starch (amylose–amylopectin complex) or the structure of conjugates with peptide or proteins (glycosaminoglycans and mucins).

1.1.3 POLYSACCHARIDE POLYDISPERSITY: MOLECULAR WEIGHT AVERAGES

This is an important concept to understand when describing polysaccharides since, whereas the synthesis of proteins is precisely regulated from the gene sequence with the result that they have precisely defined molecular weights (e.g. ribonuclease has an M = 13,683 Da), the synthesis of polysaccharides is governed by enzymes operating without a template which yield a spectrum of saccharide chain lengths. Therefore, we have to specify the molecular weight of a particular polysaccharide preparation as an average or as a distribution. There is nothing difficult about average molecular weights, but the student needs to appreciate that there are different averages depending on the definition and the technique used to measure it. The type of average specified can be an average, for example with respect to number of molecules with specified M values in the distribution (to give the 'number average', M_n) or with respect to concentrations of molecules with specified M values (to give the 'weight average', M_w). These different types of averages can be measured with different techniques which will be considered in the following. Another average which can be measured is the 'z-average' (M_z): the student needs to remember that

(1.1) $$M_z \geq M_w \geq M_n$$

where the equality only holds for a perfectly monodisperse material – an impossible practicality with polysaccharides. The various averages are inter-related by the equations

(1.2) $M_n = \sum n_i M_i \big/ \sum n_i$; $M_w = \sum n_i M_i^2 \big/ \sum n_i M_i$; $M_z = \sum n_i M_i^3 \big/ \sum n_i M_i^2$

n_i is the number of molecules of each species i whose molecular weight is M_i. Instead of the number of molecules n_i in Equation (1.2), the molar concentration c_i (mol L^{-1}) could be used since this is simply the number of molecules (L^{-1}) divided by Avogadro's number (6.022 045 × 10^{23}). $c_i = C_i/M_i$ where C_i is the weight concentration (in g L^{-1}) of each species i. The formula in Equation (1.2) for M_w therefore can be written $M_w = \Sigma C_i M_i/\Sigma C_i$ and relates directly to those techniques such as light scattering and sedimentation equilibrium (Section 1.1.5) which depend on the weight concentration of each species, whereas the formula for M_n relates directly to those techniques, which depend directly on the number of molecules such as osmotic pressure.

The ratios $I_w = M_z/M_w$, or more usually $I_n = M_w/M_n$, are used by commercial manufacturers of polysaccharides as indices of polydispersity, and these can be related to the standard deviation of the distribution of molecular weights in a preparation, whatever form this distribution takes (Gaussian, log normal etc.). Alternatively, the distribution can be measured directly as considered below using a light-scattering photometer coupled to a size-exclusion chromatography (SEC) system.

All this information is important since the functional properties of polysaccharides, besides being affected by the chemical composition and conformation, can be largely influenced by the molecular weight distribution. Although mass spectrometry, the most precise molecular weight probe (to Da precision), can, in principle, provide this molecular weight information, its application is limited for polysaccharides, because of their large size and polydispersity. However, a number of solution methods are available for providing this information.

1.1.4 SOLUTION PROPERTIES AND CHARACTERIZATION METHODS

The functional properties of polysaccharides are governed by their chemical nature (primary sequence) and their physico-chemical characteristics. Although some structural and storage polysaccharides function in the solid state (and indeed, like cellulose, chitin and amylopectin are insoluble in aqueous solvents), most function in aqueous environments ranging in concentration from dilute solution to a gel. Since it is generally in an aqueous environment that polysaccharides are used in biotechnology, it is worthwhile to briefly consider solution properties of polysaccharides and the methods used to obtain this information.

1.1.5 MEASUREMENT OF MOLECULAR WEIGHT AVERAGES AND MOLECULAR WEIGHT DISTRIBUTION

Molecular weight is one of the most important physical properties underpinning polysaccharide behaviour. For example, the ability of a polysaccharide to gel is strongly affected by its (average) molecular weight and its distribution of molecular weights.

The number-average molecular weight, M_n, for example can be measured by osmotic pressure, provided that a suitable semi-permeable membrane can be found which is impermeable to all the polysaccharide chains present in a sample (including the low molecular weight end of the polysaccharide distribution). This technique has an upper limit of M_n ~100,000 Da. A plot of Π/RTC versus C, where Π is the osmotic pressure (dyn/cm^{-2}), T is the temperature (in kelvin, K), R is the gas constant (8.314×10^7 erg mol K^{-1}) and C is the concentration (g mL^{-1}), yields the reciprocal of M_n (g mol^{-1} = Da) from the intercept and the second virial coefficient, B (mL mol g^{-2}), from the slope. Because of the peculiarity that the Dalton is equivalent to the centimetre–gram–second (cgs) unit of 'molar mass' (g mol^{-1}) and not the SI unit (kg mol^{-1}), polysaccharide chemists and physical biochemists as a whole prefer the use of the cgs system (consider the molecular weight of water is 18 Da, 18 g mol^{-1} but 0.018 kg mol^{-1}).

The weight-average molecular weight, M_w, can be measured (in solution) by one of the three principal methods: One of these is 'static light scattering' whereby the 'Rayleigh ratio' R_θ (a measure of the ratio of the intensity of light scattered by a macromolecular solution at an angle θ to that of the incident radiation) is recorded as a function of θ. Modern photometers permit the simultaneous measurement of R_θ for a range of angles. Extrapolations to zero angle (to eliminate contributions from shape) and zero concentration (to eliminate contributions from thermodynamic non-ideality) using the so-called 'Zimm plot' of KC/R_θ versus $kC + \sin^2(\theta/2)$ (where K is an experimental constant depending on the wavelength of the light used and the 'refractive increment' of the polysaccharide – generally in the region 0.14–0.16 mL g^{-1} for polysaccharides – and k is a positive or negative number arbitrarily chosen to conveniently scale the abscissa) yields the reciprocal of M_w from the common intercept, twice the second virial coefficient (represented by the symbol A_2 or B and with units mL mol g^{-2}) from the slope of the $\theta = 0$ line (after allowance for the factor k), and the highly useful conformation parameter R_g (the 'radius of gyration', cm), by a manipulation of the limiting slope of the $C = 0$ line (Figure 1.7). A_2 can also be used as a measure of molecular shape. The Zimm procedure is the simplest for interpreting multi-angle, multi-concentration light-scattering data but is not the optimum representation for all polysaccharides: the interested student should refer to an article by Burchard (1992) who describes the usefulness of logarithmic and other types of manipulation of the data. On the other hand, if measurements can be performed at sufficiently low concentration, a $C = 0$ extrapolation is not necessary to obtain M_w (although A_2 can then of course not be obtained). Another variant is 'low-angle light scattering' in which a single fixed low angle is used: extrapolation to $\theta = 0$ is, therefore, assumed not necessary (although this time R_g cannot be measured). All these light-scattering-based procedures only work if solutions are scrupulously clear of supramolecular contamination – this can be a serious problem since polysaccharide solutions often contain 'microgels' or other supramolecular components.

An important development was the coupling of light-scattering photometers (with the scattering cuvette, cell or tube replaced by a flow cell) to high-pressure SEC columns. This was initially performed with low-angle photometers, and then superceded with the now highly popular SEC–multi-angle light scattering (SEC–MALS), viz. multi-angle (laser)

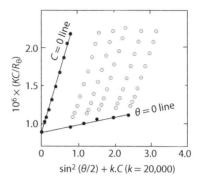

FIGURE 1.7   Light-scattering Zimm plot for a hyaluronic acid preparation in 0.15 M NaCl. K is an experimental constant (see text) and k is an arbitrary constant, in this case 20,000, chosen so as to space out the data. The common intercept on the y-axis gives the reciprocal of the weight-average molecular weight, M_w (Da or g mol⁻¹). The slope of the $\theta = 0$ line gives the second-thermodynamic virial coefficient, A_2 in mL mol g⁻², after allowance for k. The limiting slope of the $C = 0$ line gives after manipulation an estimate for the radius of gyration, R_g (cm). (Gamini et al. [1992]. Reproduced by permission of the Royal Society of Chemistry.) In cases of severe curvature, plotting the ordinate on a logarithmic scale to give a 'Guinier' plot gives a better representation (see, for example, Burchard, 1992).

light scattering coupled to SEC (Wyatt, 1992, 2013). This serves three purposes: (1) it provides an online clarification from supramolecular particles; (2) it provides directly an absolute molecular weight distribution (Figure 1.8a): the SEC separates the distribution into 'volume slices' and the MALS provides an instantaneous angular extrapolation to give the molecular weight of each 'slice' as it passes through the flow cell; because of the low concentrations after dilution by the column, concentration extrapolations are often not necessary; (3) it can provide, after considerable care in the R_g evaluation, valuable conformation information from the variation of R_g with M_w (see Figure 1.8b and Section 1.1.7). The very first application to polysaccharides was by Horton et al. (1991), who successfully used this method to characterize the molecular weight distribution of an alginate (Horton et al., 1991), and the first successful application to a glycoconjugate was in 1996 by Jumel, Fiebrig and Harding applied to gastric mucin (Jumel et al., 1996) followed shortly by application to colonic mucin. Since then it has become the method of choice for the characterization of molecular weights of polysaccharides and other glycans but there are limitations, principally (1) the requirement of a separation medium or 'matrix' – a column – which may not be inert and (2) the separation range of the columns (generally ~10⁶ Da maximum). This can be extended by using an alternative separation system – principally field-flow fractionation, which has been used to determine the molecular weights of starches of molecular weights up to 10⁸ Da (see Harding et al., 2016). This also, however, requires inertness of the matrix/membrane required for the fractionation process.

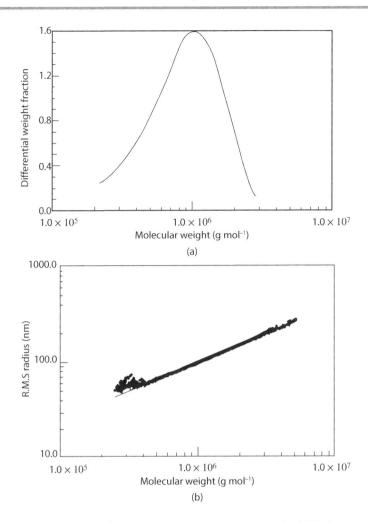

FIGURE 1.8 Size-exclusion chromatography combined with light scattering 'SEC–MALS' for determining the molecular weight distribution and gross conformation of a polysaccharide (a microbial hyaluronic acid preparation). (a) Molecular weight distribution, (b) corresponding plot of R_g versus M_w. The Mark–Houwink c (or v) coefficient from the slope is ~(0.59 ± 0.1), suggesting a coil type of conformation. (From Jumel, K. and Harding, S.E., unpublished observations.)

A second powerful and independent method for obtaining M_w and a measure of the size distribution or heterogeneity/polydispersity of a polysaccharide or glycoconjugate is analytical ultracentrifugation. Although the equipment is more expensive and can take longer to run, it is matrix free: separation and analysis without the need for a polymer separation column or membrane. In this technique, the macromolecular solution is spun (in a balanced precision-made centrifuged cell with transparent windows at the top and bottom) at high speed (up to 50,000 rev min^{-1}) and concentration distributions recorded using special optical systems based on refractometric (Rayleigh interference) properties, or, if the macromolecule has a suitable chromophore (not common for polysaccharides unless specifically labelled, but usual for glycoproteins and all nucleic acids) ultraviolet (UV)–visible absorption optics. There are two types of analytical ultracentrifuge experiment (Harding et al., 2015):

1. Sedimentation velocity

   Here a sedimenting boundary appears between solution and solvent if the ultracentrifuge is run at sufficiently high rotor speed (typically 30–50,000 rev min^{-1} for a polysaccharide of M_w ~100,000 Da). The change in the concentration distribution with time is recorded and these are transformed into distributions of sedimentation coefficient. The sedimentation coefficient (symbol s, units seconds or Svedbergs, S, where $1S = 10^{-13}$ s) is principally a measure of the size of a particle although it depends on the conformation. Distributions of the sedimentation coefficient, $g(s)$ versus s, or $c(s)$ vs s, are a measure of heterogeneity/polydispersity. In Figure 1.9 for example contrast the unimodal but broad distribution of sedimentation coefficient for an aqueous solution of an alginate with that for narrow discrete components in a solution of an aminocellulose. Distributions of sedimentation coefficient can be converted into a molecular weight distribution if the conformation is known, using a procedure known as the Extended Fujita method (Harding et al., 2011).

2. Sedimentation equilibrium

   Here, rotor speeds are chosen sufficiently low enough (typically ~10,000 rev min^{-1} for a polysaccharide of M_w ~100,000 Da) that the centrifugal force is comparable with the back force due to diffusion. Equilibrium is eventually reached between the two forces (usually >24 h) and because the final equilibrium distribution of concentration in the ultracentrifuge cell $C(r)$ versus r is stationary, frictional effects do not come into play, and hence is a function of the molecular weight and not of conformation. Analysis of the $C(r)$ versus r profile by what is called 'SEDFIT-MSTAR' analysis (Schuck et al., 2014) yields the apparent weight-average molecular weight $M_{w,app}$. Extrapolation (where appropriate) of $1/M_{w,app}$ to $C = 0$ then yields the reciprocal of the weight average M_w from the intercept and, as with static light scattering, $2B$ from the (limiting) slope. In many cases, if a sufficiently low loading concentration is employed (~0.3 mg mL^{-1}), $M_{w,app} \sim M_w$ and no extrapolation is necessary. Conversely, in cases of severe non-ideality (e.g. alginates), direct non-linear extrapolation of $M_{w,app}$ to $C = 0$ gives a safer estimate of M_w. It is also possible from a single experiment to obtain local or point average $M_{w,app}(r)$ as a function of r or $C(r)$. Extrapolation of $M_{w,app}(r)$ to $C(r) = 0$ provides a further estimate for M_w, and from further manipulations the z-average molecular weight, M_z, can be obtained.

The final popular method of obtaining molecular weight is the sedimentation–diffusion method, whereby the sedimentation coefficient s from sedimentation velocity analytical ultracentrifugation is combined

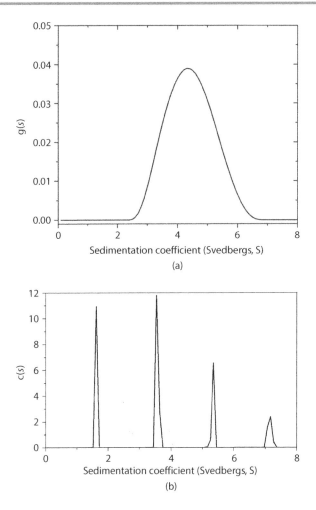

FIGURE 1.9   Sedimentation coefficient distribution plots for (a) an alginate solution, (b) an aminocellulose solution. Note the difference: the alginate has a unimodal but broad distribution of sedimentation coefficient,
whereas the aminocellulose has several clear components present. The
interested student should see Harding et al. (2015) for a more in depth
consideration.

with the diffusion coefficient, D (cm^2 s^{-1}), from dynamic light scattering
(or boundary spreading in the ultracentrifuge) via the Svedberg equation
(after both parameters have been corrected to standard conditions):

(1.3) $$M_w = (s/D)\,RT\,/\,(1-\bar{v}\rho)$$

where ρ is the solvent density and $\bar{v}$ is the partial specific volume of the
polysaccharide (usually in the range 0.5–0.6 mL g^{-1}). M_w obtained according to Equation (1.3) is often referred to as M_{SD}. For both s and D, an extrapolation to $C = 0$ is necessary to eliminate non-ideality effects. For dynamic
light-scattering measurements of D for rod-shape polysaccharides, a further extrapolation to $\theta = 0$ is also necessary: both extrapolations can be

done on a 'dynamic Zimm plot', by analogy with the Zimm plot for static light scattering – the interested student is again referred to the article by Burchard (1992). Both s and D measurements need also to correspond to the same temperature of measurement, T.

Harding et al. (2015) have provided a comprehensive review of applications of analytical ultracentrifugation to polysaccharide analysis.

1.1.6 POLYSACCHARIDE CONFORMATION IN SOLUTION

The relevant questions to be considered are: (1) What is the conformational flexibility of the polysaccharide in solution (stiff rod shape, a flexible or random coil, a compact sphere, or something in between)? (2) If it is a stiff rod, what is its axial ratio? Or, if it is a flexible coil, what is its persistence length? (3) Is there any secondary or tertiary structure (helix formation etc.) and what are its characteristics (number of residues per turn, number of chains involved etc.)?

1.1.6.1 OVERALL OR 'GROSS' CONFORMATION: HAUG TRIANGLE

For gross conformation analyses, a very simple but highly useful representation is the Haug triangle (Figure 1.10) developed in Trondheim, Norway by Arne Haug. A description of its use (in Norwegian) is given in a classical book by Smidsrød and Andresen (1979).

There are two traditional procedures for finding the gross conformation of a polysaccharide. One is the 'Mark–Houwink' (or in full 'Mark–Houwink–Kuhn–Sakurada, MHKS') approach whereby a polysaccharide is fractionated and the dependence of a hydrodynamic parameter (the intrinsic viscosity, $[\eta]$ (mL g^{-1}), the sedimentation coefficient, s, the radius of gyration, R_g, or the diffusion coefficient, D) on molecular weight for the

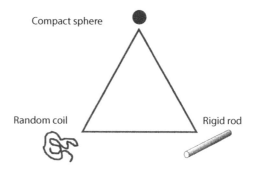

FIGURE 1.10 Haug triangle representation of polysaccharide conformation. In this representation, the three extremes of conformation are placed at the corners of the triangle. The conformation of a given macromolecule can then be represented by a locus along the sides of the triangle. Most polysaccharides will have conformations represented by loci between the random-coil and rigid-rod limits.

range of fractions yields a 'Mark–Houwink' (exponent) coefficient, a, b, c or ε according to the 'scaling' relations:

$$(1.4) \qquad [\eta] = K' M^a, \quad s = K'' M^b, \quad R_g = K''' M^c, \quad D = K'''' M^{-\varepsilon}$$

a (sometimes given as α), b, c (sometimes given as ν) and ε are usually obtained from 'double logarithmic' plots of the hydrodynamic parameter versus M, and have specific values for a given conformation (Table 1.2). The Mark–Houwink representations are particularly useful for spotting changes in polysaccharide conformation with increase in polymer chain length, and an interesting example is for amylose (Figure 1.11).

The second useful way for determining the gross conformation of a polysaccharide in solution is to use the concentration dependence or 'Grálen' (1944) term, k_s, of the sedimentation coefficient which we

TABLE 1.2   Mark–Houwink (MHKS) and Wales–van Holde Conformation Parameters

Conformation	a (or α)	b	c (or ν)	ε	$k_s/[\eta]$
Compact sphere	0	0.667	0.333	0.333	1.6
Rigid rod	1.8	0.15	1.0	0.85	<1[a]
Random coil	0.5–0.8	0.4–0.5	0.5–0.6	0.5–0.6	~1.6

a   Depends on axial ratio.

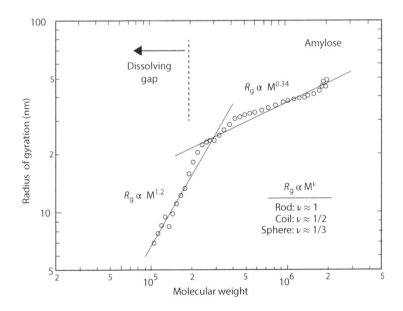

FIGURE 1.11   Mark–Houwink R_g plot showing change in conformation of amylose with increase in molecular weight. Molecular chains in this preparation seem to have a spread of conformations, and those <~200,000 Da in a more linear-rod type of conformation and those >~200,000 Da in a more spheroidal or coiled form. (Rollings [1992]. Reproduced by permission of the Royal Society of Chemistry.)

will describe after we have considered how to eliminate complications through molecular charge effects.

1.1.6.2 EFFECT OF IONIC STRENGTH OF THE SURROUNDING MEDIUM

The ionic strength I of a solution is defined by $I = \frac{1}{2}\Sigma_i c_i z_i^2$ where c_i is the molar concentration of an ion whose charge (valency) is z_i. So, a 0.1 M solution of NaCl would also have an ionic strength of 0.1 M, since it is completely dissociated into its ions. The contribution to the ionic strength of a solution from the charged polysaccharide itself is usually very small, since the molar concentration of the latter (= weight concentration/molecular weight) is usually tiny. The ionic strength of the solution has an important bearing on molecular properties of charged polysaccharides on two counts: (1) It helps to suppress (by charge shielding) the non-ideality of the system from polyelectrolyte or macromolecular charge effects: for proteins, $I = 0.1$ M is usually sufficient; for the highly charged polyanionic polysaccharides, an I between 0.3 and 0.5 M is usually sufficient; (2) The conformation of a charged polysaccharide itself will be affected: a 'stiffness parameter' B (Smidsrød and Haug, 1971) has been defined (as a warning to the student, this is not the same as the B often used to represent the thermodynamic second virial coefficient considered earlier) which gives a useful representation of the effect of ionic strength on a polyelectrolyte chain. The student can find a table of values of this parameter in Lapasin and Pricl (1995, pp. 279–281).

1.1.6.3 INTRINSIC VISCOSITY

We have already mentioned in Section 1.1.5 how M, s, R_g and D can be extracted. The intrinsic viscosity $[\eta]$ is obtained from the intercept of a plot of the reduced specific viscosity η_{red} (= $(\eta_{rel} - 1)/C$) versus C, where η_{rel} is the ratio of the viscosity of the polysaccharide solution to that of the solvent in which it is dissolved (Harding, 1997, 2013). It is popularly measured using a thermostatically controlled rotational viscometer, which permits extrapolation to zero shear (in cases where the shear dependence is negligible, simple capillary viscometers can be used). Differential pressure viscometers, which work on a different principle (and with negligible shear effects) can operate at low concentration and can be incorporated online to SEC or SEC–MALS or MALS devices, are becoming increasingly popular. These can provide simultaneously the intrinsic viscosity $[\eta](V_e)$ and weight-average molecular weight $M_w(V_e)$ both as a function of elution volume V_e. Another useful parameter for determining the overall conformation of a polysaccharide is the 'Wales–van Holde' ratio $k_s/[\eta]$, where k_s (mL g^{-1}) is the concentration dependence or 'Gralen' parameter of the sedimentation coefficient. Assuming no self-association and the absence of any intermolecular electrostatic repulsive forces (this situation can be achieved even for charged polysaccharides by 'polyelectrolyte suppression' using a salt solution of sufficient ionic strength) $k_s/[\eta]$ is also able to distinguish between rods ($k_s/[\eta]$ ~0.2–0.5) from random coils and spheres ($k_s/[\eta]$ ~1.6).

1.1.6.4 CONFORMATION ZONES

In terms of gross conformation, an alternative to the Haug triangle representation of polysaccharides is the use (Pavlov et al., 1997) of conformation types or *zones* (Figure 1.12) ranging over:

- Rigid rod (zone A), such as the triple chain helical polysaccharide schizophyllan
- Rod with limited flexibility (zone B), such as the double helical xanthan (n.b. some xanthans fall into the zone A category)
- Semi-flexible coil (zone C) such as the neutral polysaccharide methylcellulose, the polyanionic pectin (highly de-esterified) and alginate and the polycationic chitosan
- Random coil (zone D) such as the neutral polysaccharides pullulan and dextran
- Globular or highly branched (zone E) such as the neutral polysaccharide amylopectin (also neutral)

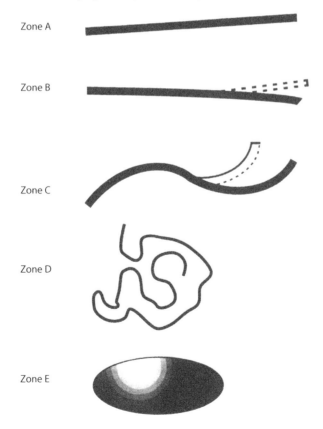

Zone A

Zone B

Zone C

Zone D

Zone E

FIGURE 1.12 Conformation zones of polysaccharides. A: rigid rod; B: rod with limited flexibility; C: semi-flexible coil; D: random coil; E: globular or highly branched.

The conformation zone for a polysaccharide can be determined using the Mark–Houwink scaling relations, but this requires fractionation of the polysaccharide, although this requirement can be avoided if the mass per unit length of the polysaccharide M_L is known (Pavlov et al., 1997). M_L can be measured by, for example, x-ray diffraction (Alexander, 1979) or electron microscopy (Stokke and Elgsaeter, 1991) (Table 1.3).

Table 1.4 summarizes the conformation types of a range of polysaccharides based on hydrodynamic and light-scattering measurements.

1.1.6.5 MORE SPECIFIC CONFORMATION INFORMATION: RIGID STRUCTURES

Once the conformation type (in terms of its zone) of a polysaccharide has been established, we can by appropriate further measurements provide more quantitative information. For example, if the polysaccharide is a rod (e.g. xanthan), it is possible to characterize it further in terms of its dimensions or axial ratio. In terms of the radius of gyration R_g (from light scattering) the length L of a very long rod is given simply by (van Holde, 1985)

$$(1.5) \qquad\qquad L = 3.464 R_g$$

TABLE 1.3   Mass per Unit Lengths, M_L, and Persistence Lengths, L_p, of Some Glycopolymers.

Polysaccharide	M_L (g mol^{-1} nm^{-1})	L_p (nm)
Schizophyllan	2150	150–200
Xanthan	1700–2000	100
Cellulose nitrate	550	40
Methylcellulose	360	15
Amylose	1400	2–4
Pullulan	340	2

Note: The list reading down is given in order of decreasing chain rigidity

TABLE 1.4   Conformation Zones of Polysaccharides

Conformation Zone	Nature	Examples
A	Rigid rod	Schizophyllan, scleroglucan
B	Rigid rod with limited flexibility	Xanthan
C	Semi-flexible coil or asymmetric coil	Alginate, pectin, chitosan, xylans, cellulose nitrate, methylcellulose
D	Random coil	Pullulan, dextran, guar, locust-bean gum, konjac mannan
E	Globular or highly branched	Amylopectin

Formulae exist relating R_g and other hydrodynamic data to the axial ratio of rods and other rigid-structure models for polysaccharide conformation such as ellipsoids and bead models, but this is beyond the scope of this book.

1.1.6.6 FLEXIBLE STRUCTURES: WORM-LIKE COIL AND PERSISTENCE LENGTH

If a linear polysaccharide is not a rigid structure then a most useful representation is in terms of a worm-like coil which covers the range of flexibilities from a stiff rod to a completely random coil. The extent of flexibility can be represented by the ratio of the contour length, L, to the 'persistence length', L_p (Figure 1.13), or just simply in terms of the persistence length, L_p, by itself (Table 1.3). Although the theoretical limits are $L_p = 0$ (random coil) and infinity (rigid rod), practically the range is from ~1–2 nm (random coil) to ~200 nm (rigid rod). By analogy with the Mark–Houwink representations, L_p can be modelled from the dependencies of $[\eta]$, s, D or R_g on M (Figure 1.14).

1.1.6.7 ORDERED STRUCTURE

Several commercially important polysaccharides such as xanthan, κ-carrageenan, schizophyllan and agarose are known to possess secondary structure (helices) under certain conditions. This will affect the optical activity of the polysaccharide: this in turn is manifested by the ability of a substance to absorb light at certain wavelengths depending on the plane of polarization of that light (a phenomenon known as circular dichroism [CD]) and also will rotate the plane of polarization at any wavelength (optical rotatory dispersion [ORD]). These methods have been applied for example to the microbial polysaccharide xanthan to show the ordered (helical)–disorder transition between low and high temperatures.

1.1.7 IMAGING OF POLYSACCHARIDES: X-RAY FIBRE DIFFRACTION, ELECTRON MICROSCOPY AND ATOMIC FORCE MICROSCOPY

X-ray diffraction of polysaccharides is not as clear-cut as x-ray diffraction studies of protein crystals, where the intricacies of molecular structure

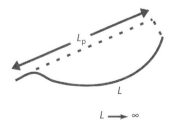

FIGURE 1.13 Length parameters for a polysaccharide: L = contour length; L_p = persistence length, defined as the projection length along the initial direction of a chain of length L and in the limit of $L \rightarrow \infty$.

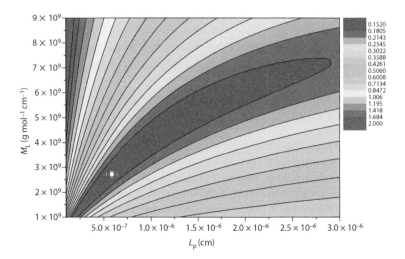

FIGURE 1.14   Persistence length determination for the capsular *Streptococcal* polysaccharide SP(4) in 0.1 M phosphate–chloride buffer. The plot known as a 'HYDFIT' plot is generated from combining relations linking data sets of intrinsic viscosity against molecular weight (measured using an SEC-MALS device coupled to a viscometer) and sedimentation coefficient against molecular weight (measured using an analytical ultracentrifuge). Simultaneous fitting to these relations involves finding the values for the persistence length L_p and mass per unit length M_L corresponding to the minimum value of a target (error) function. For SP(4) the plot yields $L_p \sim 6.2$ nm and $M_L \sim 292$ g mol^{-1} nm^{-1}. (Reproduced from Harding et al. (2012), with permission from Elsevier.)

for very many are known to Ångström (0.1 nm) precision. This problem with polysaccharides is a consequence of polydispersity, flexibility and crystallizability. X-ray diffraction from polysaccharides in fibre form has, however, provided useful information on helical content and some information about helix geometry (pitch, residues per turn, axial rise per residue) and the conformation state of the pyranose or furanose ring structure in homopolymers such as polymannuronate and polyguluronate alginates (Atkins et al., 1970). Ambiguities in data interpretation can arrive from variable alignment of polysaccharide chains and the presence of crystalline structure and this has led to differences in opinion on the secondary structure of some helical polysaccharides: an example has been the debate over whether the ordered forms of xanthan and the marine polysaccharides κ- and ι-carrageenan and agarose are in single- or double-helical form. Some polysaccharides contain a significant crystalline content and x-ray diffraction has been used to ascertain the degree of crystallinity of starch granules. In fact, starch provides a good case study of x-ray methods and the interested reader is referred to an article by Clark (1994).

Electron microscopy provides us with a direct visualization of polysaccharide structure but at a price: the environment a molecule finds itself in for analysis is even more alien (compared with its natural environment) than that required for x-ray diffraction. High shearing and surface tension forces on air-drying onto a surface prior to exposure to a high vacuum are hardly kind to molecular integrity. Alternative methods of preparation are also harsh (e.g. treatment with highly concentrated glycerol or freezing). The importance of comparing data from different preparation techniques has been stressed (Stokke and Elgsaeter, 1991). Advantage can actually be taken within transmission electron microscopy (TEM) of the high shearing and surface tension forces for unscrambling the contour length of polysaccharides, enabling determination of the mass per unit length M_L – a highly useful parameter as noted earlier. This technique has also proved useful for examining order–disorder transitions and association and gelation characteristics (branching points or junction zones etc.). Scanning electron microscopy, which scans the surface of an object, has been particularly useful for studying particulate and composite gels like starch (Hermansson and Langton, 1994).

Atomic force microscopy also provides us with direct visualization whilst keeping the polysaccharide in direct contact with solution. It is not a true solution technique in that the macromolecule still has to be immobilized onto a surface, but at least it is closer to the real situation for many glycans. A highly sensitive detector profiles the surface of the macromolecule – this method has, for example, been very successfully used by Deacon et al. (2000) for the study of chitosans and their mucoadhesive interactions with mucins (Figure 1.15).

1.1.8 POLYSACCHARIDE RHEOLOGY

This has been of considerable interest, particularly in the area of food technology. Whereas hydrodynamics refers more to dilute solution properties of polysaccharides, rheology is associated more with highly concentrated dispersions (colloids, gels, emulsions etc.), and a wide variety of viscometers or rheometers, and even the ultracentrifuge, have been used to study these properties. The classical 'transition' between what the rheologists term 'dilute' and 'semi-dilute' behaviour is represented by a concentration parameter 'C^*' also known as the 'total occupancy concentration', since at this concentration molecular domains (including a large layer of chemically and physically associated solvent called 'hydration') are considered to interpenetrate. (C^* is not the same as the critical gel concentration, C_0.) C^* is considered to vary as M/R_g^3 and the following approximation has also been useful (Launay et al., 1986):

(1.6) $$C^* = 3.3/[\eta]$$

Equation (1.6) has, for example, been found to fit viscosity, sedimentation and molecular weight data for xanthan very well (Dhami et al., 1995). Other useful parameters when describing rheological characteristics of polysaccharide dispersions have been the shear modulus G (N m^{-2} in SI units or dyn cm^{-2} in the cgs system) and for viscoelastic materials the storage shear modulus G', the loss shear modulus G'' and the complex shear modulus $G^* = (G'^2 + G''^2)^{1/2}$.

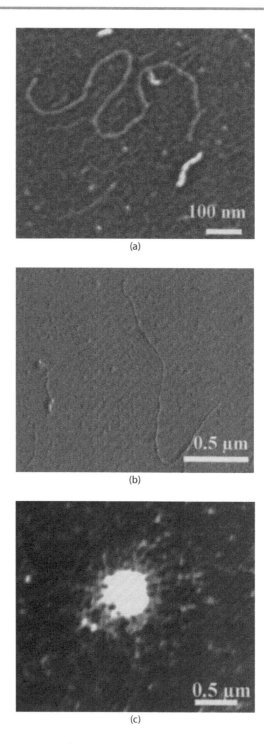

FIGURE 1.15   Atomic force microscopy images for a chitosan (a), mucin glycoprotein (b) and the complex between chitosan and mucin (c). Macromolecules and complex immobilized to a surface but in contact with 0.1 M sodium acetate buffer. (Reproduced from Deacon, M.P. et al., *Biochemical Journal*, 348, 557–563, 2000. With permission.)

1.1.9 CLASSIFICATION OF POLYSACCHARIDES

There is no single method of classification which serves all purposes. Polysaccharides could be classified purely on the basis of their chemical structure, and in chemistry textbooks they will be found organized in this way. However, because this book is about their biotechnology we have chosen to classify them with more emphasis on their biological source. It would also have been valid to classify them on the basis of their use, which would have found some bacterial polysaccharides alongside gums obtained from seeds, since they have the same uses. We have cross-referred where appropriate, but have treated bacterial polysaccharides separately. This is justified because at least part of the potential of biotechnology lies in the manipulation of biosynthetic pathways, which relates directly to the source organisms.

We have also classified plant polysaccharides on the basis of function. This can be justified on the same grounds, but also because their properties and to some extent their utility is linked to their function in the source plants. Thus, structural polysaccharides, in effect those of the cell wall, are different from the storage polysaccharides. Mammalian enzymes do not break down the components of the cell wall. Ruminants have to rely on microorganisms held in specialized compartments of their gut to do this. On the other hand, most of the storage carbohydrates, found predominantly in seeds, are broken down. This tells us something about the evolution of the mammalian digestive system, and the way in which mankind has selected his food plants, and plants for other purposes, and the way in which they too have evolved. Because some of the storage polysaccharides, for example arabinans, are not broken down and can cause problems as a result, potential improvements to raw materials can also be seen in context.

A general account of cell wall structure and storage carbohydrates is given in this chapter and much more detailed accounts will be found, where they are relevant to the biotechnology under discussion, in the following chapters. Some topics do not fall easily into this scheme and for example the use of isolated enzymes to synthesize polysaccharides will be found in the section on bacterial systems simply because many of the enzymes used to come from microorganisms. We have taken fungi and algae to be plants, for the purposes of cell wall components and storage carbohydrates, but references to fungal enzymes will be found throughout where their use is considered. This chapter contains a brief consideration of the use of fungi and bacteria as producers of enzymes useful in altering polysaccharides and the processes involved.

1.1.10 THE CELL WALL

The components of the cell wall will be very important targets for biotechnology, and there is already work aimed at such diverse uses as paper-making, cotton filaments in textiles and tomato puree manufacture. The consumption of wood in all its forms is enormous. In the United Kingdom, it amounts to one tonne per head per year.

Cell walls form during the growth of the cell, and are initially fairly thin and able to resist turgor pressure of at least 30 atm. They are flexible, but become more rigid as the cell reaches its limiting size. In large plants such as trees they become lignified and able to support the gravitational stresses on branches as well as maintain the structural integrity of phloem and xylem. These changes are associated with changes in the composition of the cell wall, which has a complex lamellar structure and, even in simple plants, many different components. They are considered in detail in subsequent chapters, but a summary of the main features is included here as part of a consideration of the main thrust of biotechnology developments. There is considerable variation in cell walls over the whole range of the plant kingdom.

1.1.10.1 CELLULOSE

Cellulose, a (1→4) linked β-D-glucan, is an invariable component of higher plant cell walls. It forms the main fibrillar structure, though it is not always the major component. In cereal endosperm cells it can be as little as 4%, while in cotton seed fibres it rises to 94%. It is the most abundant biopolymer on earth, with an estimated 10^{12} tonnes present at any one time and an annual production of ~10^{11} tonnes. It is insoluble in water, though fragments of the chains may be soluble, which suggests that its ability to form an ordered packed structure stabilized by hydrogen bonding to give recognizable fibrils is the reason for this.

1.1.10.2 MANNANS

In certain seeds, as well as the walls of red and green algae, the fibrils may be made up of mannans, which comprise (1→4) linked β-D-mannan where both crystalline and low molecular weight forms have been described. It appears to substitute for cellulose in its general function.

1.1.10.3 CALLOSE

Commonly found in wound repairs, this (1→3) linked β-D-glucan is associated with cellulose, and may even contain some 1→4 links as well. It is present in the walls of fungi and generally seems to be analogous to cellulose in its function though perhaps capable of a more transient existence. In contrast to callose, *in vivo* degradation of cellulose is virtually absent, though disappearance of cellulose fibrils during fruit ripening has been reported in pears, apples, avocados and peaches so it may be a general phenomenon. Other cell wall components become easier to extract at the same time. The kernel of the date has very thick cell walls and is very rich in mannans. Here, however, they are mobilized during germination and must be regarded as having a dual function including storage.

1.1.10.4 HETEROPOLYMERS

Other glucans with 1→3, 1→4 and 1→6 links in various distributions, either block or random, have been described, and again fill a cellulose-like function in the walls. They are, however, mobilized during seed germination. All these materials, despite their varied linkages, behave as single unbranched chains, for the most part capable of forming helical structures involving many chains in side-to-side association. The glucans of barley have been much studied because of their importance in brewing. The responsible enzyme systems occur as a complex.

1.1.10.5 HETEROGLUCANS

These, also known as xyloglucans (or sometimes 'amyloids'), were first found in the tamarind (*Tamarindus indica*) and are essentially cellulose with side chains. The first residue of the side chain is always a xylopyranose residue, and this may carry further residues including arabinose, fucose and galactose. Clearly the side chains modify its properties and its role in the wall structure is unclear: they certainly interact with both the cellulose and the pectin, and covalent links may be involved with the latter.

1.1.10.6 HETEROXYLANS

These are chains of (1→4) linked β-D-xylopyranosyl residues, with extensive side chains, and the main chain can also be acetylated. They may also carry phenols. These too make up part of the fibrous structure of the wall.

1.1.10.7 GALACTOMANNANS, GLUCOMANNANS AND GALACTOGLUCOMANNANS

These, as their names suggest, contain galactose residues, and glucose residues attached as side chains to the main mannan chain. The proportions are high, reaching a ratio of 1:1:3 galactose:glucose:mannose with DP ranging up to 4000. They are generally characterized as *gums* when obtained from certain seeds. They tend to be at high levels in the walls of conifers, reaching 15%. Their solubility depends on size and degree of side chain, and they are of interest as gel formers and thickeners. While they occur in the cell walls, they are likely to have a storage function as well and can be formed and degraded depending on the physiological status of the plant. In seeds of the *Liliaceae* and *Iridaceae* and in lupins (*Lupinus angustifolias*) they are mobilized on germination, apparently from very thick cell walls. They do not occur in the seeds of dicotyledons. Breakdown is a hydrolytic process and suitable enzymatic activities have been found. Little is known about the enzymes, and almost nothing about biosynthesis. There has been some dispute about whether they are storage components or cell wall components. Clearly they are both, but whether they are part of the cell wall matrix, or merely function as a sort of filler, is not yet resolved.

1.1.10.8 PECTINS

These are the most complex polysaccharides found in the cell wall and contain main chains which are composed of residues of galacturonic acid, mostly methylated, with rhamnose residues at intervals in the chain. The side chains contain arabinans and galactans, may also have a few phenolic ferulic acid residues, and are probably covalently linked to other cell wall components. Pectins can be the major component of cell walls in fruits.

1.1.10.9 GLYCOPROTEINS OR EXTENSINS

These are probably associated with fresh wall synthesis, and are often seen as a result of wound healing. They are hydroxyproline-rich proteins, extensively glycosylated on the hydroxyproline residues, and contain up to 50% carbohydrate. They have a rod-shaped structure, and a cDNA corresponding to a 306 residue chain has been obtained from carrot. The protein part bears an obvious resemblance to gelatin, and its parent collagen, and as such is of very considerable interest to biotechnology. A 'vegetarian gelatin' is currently seen as a highly desirable target. At least some has been found in the seed coat of the soybean. The extensin appears to be cross-linked through phenyl–ether bridges, but not covalently to the polysaccharide components. There is also a report of a glycine rich glycoprotein, as well as sundry other proteins present in small amounts.

1.1.10.10 LIGNINS

Although these are not polysaccharides – they have a polyphenolic structure – they have a major influence on utility of all the other cell wall components, if only by making them very difficult to extract. Modification of lignins may in some cases be central to the uses of polysaccharides. Both adventitious and induced mutants affecting lignin are known in maize, sorghum and rice, and at least one disease, rubbery wood in apples, shows abnormal lignin formation. Lignins tend to be covalently linked to polysaccharides in the wall, though not to cellulose.

1.1.10.11 ALGINATES

These are found in the brown algae, and are block copolymers of guluronic acid and mannuronic residues interspersed with alternating residues. They are linear, with DP up to 1000, and are clearly similar to pectin in general function in the cell wall. They are used as sodium or potassium salts, and can be gelled by addition of calcium ions. They are often used in the form of propylene glycol esters, produced by chemical reaction with propylene oxide. Alginates have also been found in bacteria, though not approved for food use.

1.1.10.12 CARRAGEENANS

These are also found in algae, this time the red species, and are unique as the only sulphated polysaccharides in general use. They are polymers of galactose and anhydrogalactose residues, and residues may carry one or two sulphate groups. Because of the numerous strong acid groups they can interact especially strongly with proteins, and can also have synergistic effects with other stabilizing neutral polysaccharides.

1.1.10.13 FURCELLARAN

This is a name used for a very similar material obtained from *Furcellaria fastigiata* around the coasts of Sweden and Denmark.

1.1.10.14 FUCANS

These are also sulphated polysaccharides, made up of residues of fucose, xylose and glucuronic acid, found in the *Phaeophyceae*. Their resemblance to heparin has led to some interest in possible pharmaceutical applications, but no food uses are envisaged.

1.1.10.15 OTHER COMPONENTS

There are numerous other components, mostly lipids such as cutin, waxes and a carotenoid polymer found in pollen grain. The more interesting one is silica, found widely although possibly best known in diatoms.

1.1.11 STORAGE POLYSACCHARIDES

As will be evident from Section 1.1.9, while it is possible to make a clear-cut distinction between, say, cellulose cell walls and starch grains in higher plants, in many species the distinction is not so sharp. In seeds, for example, virtually any of the cell wall components might eventually be mobilized to aid the growth of the new plant so many of the components considered below might have been included as cell wall components simply because of their location. Animals have storage polysaccharides, and though they do not have specialized cell wall polysaccharides they do have carbohydrate-based structural materials like the chondroitins.

1.1.11.1 STARCH AND GLYCOGENS: α-GLUCANS

Starch is a major component of the human diet, as well as being the most important raw material for further processing (Xu et al., 2014). Glycogen is the most important storage polysaccharide found in animals, and is widespread. In mammals, it occurs in the liver and musculature. The distinction between glycogen and amylopectin is not absolute, and mainly based on its origin. α-Glucans show species variation, and some that are found in algae are close to glycogen in overall structure. Starch usually occurs as an insoluble mixture of two forms, as grains with considerable radial structure. Amylose, amylopectin and glycogen are all built from α-D-glucose residues through 1→4 links, but differ in the degree of branching. Amylose is unbranched, with a DP of about 1000 though it may have a few 1→6 links in the chain, while amylopectin is branched through 1→6 linkages at about every 20–25 residues and has a DP of around 1 million. Glycogen is more highly branched, at every tenth residue, and has a DP of about 100,000. Thus, they are very large molecules of low solubility. A variety of enzymes are known which degrade them by heterogeneous attack on the insoluble substrate. Because of the branching at least two enzymes are needed to bring about complete hydrolysis, and enzymatic hydrolysis is a fertile area for biotechnology.

The red algae produce a so-called floridean starch which occurs in granules, but behaves like glycogen when treated with hydrolytic enzymes, and appears to contain a high proportion of 1→6 linkages. The green seaweeds contain both amylose- and amylopectin-like materials, though not at high levels. About 1% seems typical, and as things stand they are unlikely to become a source of starch. Yeasts contain glycogen-type storage carbohydrates, including *Saccharomyces cerevisiae*, and some strains can accumulate large amounts. It appears to have a connection to the ability to flocculate, a property of practical importance in yeast applications.

Some α(1→3) and α(1→4) linked dextran-like materials have been described in *Aspergillus*, where they are located in the cell wall and can reach high levels (30%), though these may be mutants lacking the degradation enzymes. Something similar has been found in lichens.

1.1.11.2 FRUCTANS

These are polymers of fructose and are widespread in higher plants and algae, though they occur at high levels in very few, where they substitute for starch. They are of interest as a source of fructose, and because their synthesis is based on the transfer of a fructosyl residue from sucrose, rather than by an nucleoside triphosphate (NTP) activation mechanism. Mechanisms involving nucleotides tend to be difficult to exploit in biotechnology.

1.1.11.3 β-GLUCANS

These are widespread in both higher plants and algae. Laminarin is one of the best characterized, being found in many *Laminaria* species. It has a main chain with β(1→3) linkages, but branches through 1→6 also occur. Another well-characterized example is paramylon, from the *Euglena* species, and chrysolaminarin from diatoms. They are also widely found in fungi where they have a storage function.

In higher plants, barley glucans have been much studied because they are significant in brewing. Detailed studies of their behaviour on germination show that while they are located in the cell wall, they are rapidly broken down, and do have a storage function (see Section 1.1.10.4). On the other hand, starch grain mobilization and cell wall destruction proceed together. Other cereals also have glucans in the cell wall.

1.1.11.4 GALACTANS AND ARABINOGALACTANS

Pure galactans are rare, but have been found in potato (*Solanum tuberosum*). They are β(1→4) linked. There is a fairly important arabinogalactan in soybeans where it is one reason for the extremely high biochemical oxygen demand (BOD) of residues from soybean processing. They probably also cause similar problems in other oilseeds.

1.1.11.5 GUMS

Many gums appear to be located in the cell wall where they probably have relatively little structural function though they may have a part to play in resisting microbial invasion. On germination they tend to behave like the other storage carbohydrates and are mobilized rapidly.

1.1.12 POLYSACCHARIDE IDENTIFICATION OR FINGERPRINTING

Not only is biotechnology often concerned with producing the right blend or mix of polysaccharides – the classical examples are mixing galacto-mannans with xanthan, carrageenans or agarose to give favourable gela-tion characteristics, but also the reverse problem can be important: what is in a given polysaccharide preparation? How pure is a given sample? This latter feature is often important in biomedical applications (e.g. dextrans as plasma expanders) and pharmaceutical applications (e.g. xanthans, alginates or chitosans in slow release or bioadhesive formulations). Purity from protein or nucleic acids is relatively easy to detect. This can be done by simple spectrophotometric assay at 280 nm (for protein) or 256 nm (for nucleic acids), density gradient analytical ultracentrifugation (of the type used in the classical demonstration to show that replication of DNA is semi-conservative) or (for proteins) the use of sodium dodecyl sul-phate polyacrylamide gel electrophoresis (SDS-PAGE). Identification of an unknown polysaccharide, or the presence of unknown polysaccharide impurities, is, however, very difficult (Baird, 1993), largely because of the physical polydispersity of polysaccharides (all cases) and also chemical heterogeneity (many cases). The strategy involves identification of charac-teristic groups: for example, uronic acids in the cases of pectins and algi-nates, pyruvate groups (xanthan), sulphate groups or anhydro residues (carrageenans and agarose). For further details, the student is referred to the review of Baird (1993) and references cited therein.

1.2 THE POTENTIAL FOR BIOTECHNOLOGY

Biotechnology has its roots in ancient empirical processes of which the most obvious is the use of yeast for making alcoholic drinks and bread. There are many others from all parts of the world, and in one sense bio-technology is not new. What has happened is that in the last 40 years, a collection of novel techniques have been developed that have radically improved the feasibility of turning ideas for new processes and products into reality. They can also improve existing processes, making them more efficient. It is, however, the ability to do things which were simply not possible a few years ago, because the necessary discoveries had not been made, and the required understanding had not emerged, that has caught the imagination. We are now beginning to see applications for these novel techniques reaching production and in a few cases the supermarket shelves. It is also becoming clear that as time passes the range of appli-cations is getting wider, and ultimately every industry with a biological raw material or basis will be affected. Since this includes agriculture and the food industry, easily the two largest occupations of mankind, as well as general health care and pharmaceuticals, arguably the industry most likely to expand in the future, its importance is obvious.

1.2.1 STRATEGIES FOR BIOTECHNOLOGY

Biotechnology can be applied at a number of points in the chain that stretches from primary production to the product for sale. However, whether it is a cocktail of enzymes designed to clear out blocked drains, or an elegant modification of an enzyme system to change the colour of peonies, there is always an application in view. There is always an outcome, be it a service or a product, which must meet someone's needs. Without this end point it is not technology. It could well be said that many of the techniques and discoveries we shall be describing in this book were not developed with any particular application in view. This is true, and there are also numerous studies which have not yet found an application, and which we will not be considering. The arguments for and against doing research without a product in mind, on the grounds that virtually all research in this field will one day find some application somewhere, will no doubt rage on for many years. In this account, we are concerned only with projects that do have something that will eventually be offered for sale as an outcome. This does not mean, however, that we have restricted ourselves to examples where the product or process is well established. In many cases, it is not certain that the proposed piece of biotechnology will be a commercial success. The field has not been developing for long enough to know what the chances of success are. We know, for example that only about one-third of the new products that appear on supermarket shelves actually succeed. The success rate for biotechnology products appears to be quite high, although there have been one or two spectacular failures amongst products that actually reached the market. There have certainly been many more that never quite made it that far, and were seen to have little chance before testing the public propensity to buy.

1.2.2 USE OF ENZYMES IN PROCESSES: GENERAL CONSIDERATIONS

A large proportion of modern day biotechnology is about enzymes, not just in the academic sense but in making use of them to carry out chemical reactions. Enzymes were in use long before the development of the chemical industry which has to some extent displaced them, though in recent years they have begun to recover lost ground. Traditional biotechnology makes use of changes brought about by the use of whole organisms, though there can be little doubt that they are due to the work of enzymes contained within the cells. They can indeed be regarded as packages of enzymes, and over the centuries they have been selected for packages which are well suited to the job they must perform. This is not necessarily a simple one. Yeast is needed primarily for its ability to convert sucrose and a variety of other sugars to ethanol, but the moulds responsible for the development of flavour in cheeses carry out a wider range of reactions. There are limits too to the results of selection; no yeast could use lactose or produce starch degrading enzymes. The tendency over the years has been to use first fragments of organisms and then more or less isolated enzymes to carry out reactions.

There are, however, limitations to the availability of enzymes. Even if an enzyme is known which is able to carry out the required reaction, there is no certainty that it is available in the quantities needed at a reasonable cost. It is also quite common to find that the enzyme is found in an organism that cannot be used for safety reasons. In practice, the majority of the enzymes that have found applications in industry are straightforward hydrolases which are available from fungal or bacterial culture media. They are used in a relatively crude form, and the specified enzyme often forms less than 1% of the protein content. It is sometimes necessary to remember that while a commercial enzyme sample will contain the enzyme specified in the contract for purchase, invariably specified on the basis of activity in some mutually agreed test, there is no guarantee that other enzymes are absent. Absence as well as presence must be specified since, in practice, other enzymes will be present. There is a rather good example in Chapter 2, where an enzyme needed to degrade pectin in a specific way could not easily be obtained free from other enzymes which also degraded pectin but in an undesirable way. The required enzyme, a pectin esterase, was eventually found, free from undesirable activities, but itself a contaminant of papain, a protease used for quite different purposes.

More recently, much work has been done on methods for immobilizing enzymes so that they can easily be used in continuous processes. Despite the huge amount of research in this field, which undoubtedly attracted many academic chemists and chemical engineers, there are only two significant processes that actually use immobilized enzymes. In practice, for a variety of reasons it is often better to use enzymes once only on a batch basis without recovering them. This has implications for the cost of the enzyme. There is a better case for using a recoverable, immobilized enzyme where a purified and expensive one has been chosen. Unfortunately, it is difficult to point to any process where a purified enzyme is used, and in one instance it was shown that using a purified enzyme was no advantage. This is in the interesterification of lipids in a process for making cocoa butter, where the lipase is actually immobilized in a film of water in a column support. Using a crude enzyme gave the same results as using a purified one, and it was concluded that the rate limiting factor was not the availability of the enzyme. It was probably the rate of diffusion of the substrates. Although this is nothing to do with polysaccharides it is mentioned as probably the only example where the point has been investigated.

The general enzyme classification has six classes:

1. *Oxidoreductases.* The main limitation on the use of this class is their requirement for cofactors such as NAD and NADH. While it is possible to include ancillary systems to produce these, in practice the only applications are very small scale as in diagnostics and monitors. Glucose oxidase is the best known example.

2. *Transferases.* These are of particular interest in polysaccharide synthesis, and have been exploited in making cyclic dextrans. They have considerable potential in small-scale complex polysaccharide synthesis.
3. *Hydrolases.* By far the major group, representing the largest applications. They are often serine hydrolases.
4. *Lyases.* Limited applications, one example being in the manufacture of aspartame by phenylalanine ammonia lyase.
5. *Isomerases.* Glucose isomerase is the main example.
6. *Ligases.* Of interest in DNA manipulation.

There is a surprising amount of variable terminology in the enzyme field which has its roots in the history of biochemistry. In classical biochemistry, enzymes were identified solely by their activity. In a typical case, the substrate, usually a small molecule, was added to a tissue homogenate, and if it underwent a chemical change, the reaction was tested to see if it was due to an enzyme. The test consisted simply of boiling the sample for a minute; if the activity disappeared it was due to an enzyme. The only other test that could be done was to vary the concentration of the substrate, leading straight to classical Michaelis–Menten kinetic analysis. In fact, the main metabolic pathways such as the citric acid cycle were established in this way, without any real knowledge of the enzymes involved at all, other than what they did.

It is commonplace to say that an enzyme is absent when what is really meant is that its activity is absent. When nothing was known about enzymes there was no ambiguity, but now that we know enzymes are often present but with a mutation that makes them inactive, or maybe in an inactive conformation, a more precise definition is needed. In this book, by the 'enzyme' we mean the peptide chain specified by the gene, including mutants of the gene, whether or not it is in the active conformation. This has gradually become the commoner meaning of the word, which makes it necessary to specify the activity as a separate point.

A very significant aspect of biotechnology is that it makes hitherto unavailable enzymes potentially available. They can be produced in a different and more convenient organism, at promoted high levels. It is even possible to make arrangements for high levels to be secreted into the culture medium in the case of microorganisms, thus avoiding difficult cell breaking and extraction procedures.

1.2.3 CHOOSING A PROJECT

Despite a certain amount of study in business schools, it is not easy to say why one project is chosen for investment rather than another, and it is even more difficult to say where the list of possible projects for consideration comes from. At least four sources can be identified.

1.2.3.1 INCREMENTAL IMPROVEMENTS

In an established industry, most projects are based on the development of the existing situation. Thus, a marginal and incremental approach is common. For example, a small change in the process may lead to small changes in quality, so that the product is improved, or perhaps can be made slightly more cheaply. Ingredient cost analysis is something that all research and development (R&D) departments in the food and pharmaceutical industry spend much effort on. In manufacturing operations much the same thing goes on under a variety of names.

In a few cases, security of supply may be an issue. In the field of gums, for example, many come from quite localized areas, and depend on a traditional style of agriculture which may be disappearing. This can lead to the sudden disappearance of the gum in question, or more likely to a sharp increase in price as the producers adapt to a more modern way of life. Conversely, in other parts of the world advances in agricultural practice may lead to the appearance of novel raw materials, which must be evaluated. Another source of new raw materials is the discovery in some remote part of the world (that is to say remote from the manufacturer) of a material which, while well known to the locals, has obvious potential in some other context. There are many examples in the polysaccharide field. The Japanese or the Welsh, who were using seaweed, alginates and their derivative products centuries ago, did not imagine that they would be useful in making ice cream. Sometimes, projects arise because what used to be a cottage industry is becoming larger scale and factory based. The formulation of modern manufactured food products often includes a variety of polysaccharides, sometimes a bewildering variety. One is bound to ask why six different gums should be used. Surely one would do? As we will see in this book, there are very useful effects to be obtained from mixtures of different types of polysaccharides, but one reason is that supplies are uncertain. The ideal formulation is one where a number of different components can be used in mixtures over a range of relative contents, without affecting the product. Thus, the manufacturer can go out and buy whichever happens to be available, or cheapest, in the market at the time.

1.2.3.2 NEW RAW MATERIALS AND BY-PRODUCTS

Thus, projects can arise because someone proposes a new raw material. This is often driven by the need to find an outlet for a by-product, and the classic case for this is the need to find uses for the residues left after extraction of lipids from oilseeds. Although the major oilseed, the soybean, does not contain starch, many others like groundnut do, while cereals like maize, which can be processed for oil, are also important sources of starch. Over the years outlets for these residues have been found, mostly for feeding cattle and other farm animals, with the result that the value of the residues is now roughly equal to the value of the lipids. There is still a constant search for ways of increasing the value of agricultural residues of all kinds. Xanthan gums are one example to be found in Chapter 6. In recent years, great efforts have been made to divert some of the oilseed residues into human food use, though with no great success so far. Cellulose, lignin and pectin cell wall is a prominent component of all these materials, and for the time being the only efficient converter of this is the bacteria of the rumen, hence its predominant use in cattle food. There is no shortage of proposals for using it in other ways. All this is given added impetus nowadays because disposal of organic waste has become more demanding and much more expensive, making it even more desirable to find a use for it.

1.2.3.3 INVENTIONS

Completely novel ideas are not common and when they do arise are most likely to come from research workers in the field, most often those already working in the industry. Ideas can also come from others, possibly academics engaged in research. At the present time, researchers are much more conscious of the possibility that their work might find application than hitherto, and there are regular mechanisms for review of publicly funded research with this in mind.

When such a proposal arrives the first step is to evaluate it for feasibility. This is an elusive concept which can lead to a great deal of misunderstanding because research workers, and managers in charge of developments, tend to mean rather different things by the word. There is a primary evaluation which simply asks whether the proposed process can actually be achieved at all. A proposal to breed elephants with yellow and black striped fur would probably be rejected as not feasible by anyone, no matter how much effort went into it. But, the manager would not reject it for the same reasons as the research worker. (In the present stage of development of biotechnology, he or she might well believe that it could be done.) He would reject it because it seems to him to be a completely idiotic thing to do, with no prospect of successful marketing or use, especially since his main business is in making ice cream. The research worker would reject it because he does not think it could be done, and would probably start trying to think of ways in which it might be done one day. The criteria that different groups apply need careful definition. In one early project management system, the project manager was required to estimate feasibility at intervals on a scale running from 0 to 100. It was expected that for a successful project, the estimate would rise perhaps from 10 to 90 as it became clearer that the project was about to reach a conclusion. A fall would of course be an indication that snags had occurred. This had to be withdrawn because most project managers treated feasibility as an absolute – something is either feasible or it is not – and rated their projects as 100 from the start.

Assuming that technical feasibility, which is broadly whether anyone under any circumstances could achieve it, is assured the proposal must be assessed by a whole series of other tests. Is it worth doing? Can our organization finance it? Is it even in our kind of business, or do we want to start a new one? How long will it take? A good many proposals fail on technical feasibility, especially with biological materials. One reason for this is that the laboratory procedures of molecular biology are very difficult to scale up.

Time scales are important in biotechnology because the longer it takes the more it costs. There are no quick ways to manipulation of higher plants, and only projects that are likely to yield large cost savings or lead to especially valuable products are likely to be undertaken. In the case of a polysaccharide, if the annual turnover for that component is less than £20 million it is probably not possible to justify the costs of a genetic engineering exercise. This may rule out a number of otherwise attractive possibilities with the less common gums. Polysaccharides tend to have a low unit value – a few hundred pounds per tonne is not unusual – in sharp contrast to the very high value of pharmaceuticals produced by similar methods, which means that production facilities have to be large, and use therefore large amounts of process aid enzymes. This in turn requires that the enzymes cannot be too highly priced, so much that even if they are available they cannot be used in bulk processes.

1.2.3.4 RESPONSES TO GENERAL INNOVATION

Most large companies use their R&D organization to keep a general eye on the way the relevant fields of science are changing. No one wants to be taken by surprise when a competitor brings out a new and vastly superior product based on scientific developments that had passed unnoticed in their own organization. Nevertheless, competitors do bring out new products, and this sometimes requires innovation in order to at least match the new material. Since patents may exist, or other constraints such as a shortage of the necessary raw materials, it may be obligatory to find a different way of matching the novelty. (In one example, the introduction of a fluoride toothpaste was delayed because one company had secured the entire supply of the only suitable kind of tube for several years ahead.) This kind of thing is usually known as a 'me too' operation, and provides a steady stream of work for the R&D department.

1.2.3.5 PURE INVENTION

Finally, there are now companies many of which are "Small Medium Enterprises" or "SMEs" which exist solely to do research, often of great inventiveness, which expect to get their return by licensing their discoveries to others who will undertake the manufacture and marketing operations. This is a division of labour which was almost unknown before the advent of biotechnology. It depends heavily on the protection given by the patent system. Many of these companies have been set up by universities, or groups closely associated with them, and it is becoming clear that, rather than simply buying the rights to an innovation, or accepting a royalty or licensing arrangement, some of them are simply being taken over by larger organizations with manufacturing operations. Naturally this only happens when they have been successful in bringing a new product to the point where decisions about manufacturing must be made. A reason for this appears to be that a group of research workers that have carried out one development tends to go on and invent more. Another reason is that biotechnology is so new that the easiest way for a large pharmaceutical company to acquire a biotechnology team is to take one over. This is much quicker than trying to build one from scratch.

There are other considerations: in the early days of biotechnology it became clear that the field was so specialized that it would almost always be best to go to a specialist company rather than attempting to do the operations in house. This did, however, assume that only one project was in view. The potential applications are now appearing so frequently that a large pharmaceutical company and probably companies in the agri-business field as well should be able to keep a biotechnology group fully occupied. Even now, however, if an enzyme is involved the people who originally isolated it, and obtained transgenic expression in the laboratory, will probably go to a specialist enzyme producer for actual production. They may license the producer to sell it to other users as well.

Thus, at the moment the actual operations involving biotechnology tend to be quite fragmented between different organizations, and it is doubtful if there is one anywhere which has carried out the entire operation from concept to full-scale enzyme production to placing products on the supermarket shelf.

Patents confer the right to prevent other people from operating the process described in the patent. They do not give the right to operate it, which may itself infringe other people's patents. What tends to happen, in practice, is that companies build up portfolios of patents which they then trade with others to obtain the necessary rights. It has been found in the past that it is rarely possible to recover the costs of research by selling patent rights and there are very few patents worth the substantial sums that may be needed to defend them in litigation. When a biotechnology company makes a discovery and brings it to the point where another body can manufacture and sell it, the extent of collaboration and information transfer is greater than that simply contained in the patent. This may be another reason why collaboration so often leads to merger.

1.2.4 STEPS IN BIOTECHNOLOGY: PRODUCING AN ENZYME

Some of the examples to be found in this book involve using an enzyme, and in this section we consider the various stages that must be passed in order to obtain a supply of a suitable enzyme, by heterologous expression.

1.2.4.1 ISOLATING THE ENZYME

Suppose that an enzyme with the required activity has been found, but that it has never previously been isolated. A good example is the galactosidase of guar, which was found to have a desirable activity, and was wanted in substantial amounts (see Chapter 2 for an account of this application). The first step is to isolate it, with the primary objective to obtain sufficient amino acid sequence information to allow us to screen the publicly available DNA databases. It is possible to obtain useful sequence information from as little as 2.5 µg of protein purified by gel electrophoresis. If a DNA sequence encoding the enzyme is available from a suitable search of the public nucleotide databases, the entire amino acid sequence of the protein can be readily obtained. This of course depends on there being DNA sequence information available on your organism of interest. This is why biochemical and molecular studies are usually best undertaken on an organism where the genome sequence is readily available.

In practice, the purified enzyme is needed for a few other purposes as well as sequencing. An antibody will be very useful for a variety of analytical tests. It needs about 10 mg of protein, with appropriate adjuvants, to elicit an antibody in a rabbit. This would be for a component of average antigenicity, and the outcome of immunizations can never be guaranteed. The response may fail, or more annoyingly the antibody may turn out to be against an impurity that happens to be particularly antigenic. Rabbits can yield reasonable volumes of antisera, likely to be sufficient for any possible need, including preparation of affinity columns. It is possible to obtain an immune response in a mouse with about 25 mg of protein, but the volume of serum is very small.

An alternative and previously popularly used strategy is to generate peptides from the amino acid sequence and raise antibodies to these and also generate monoclonal antibodies.

Isolation of purified enzyme from a plant tissue can be challenging depending on the levels of the specific protein in a particular tissue. To take an average instance, suppose that the enzyme is present to the extent of about 0.1% of the protein in a typical plant tissue. This will have about 20% protein, so that the amount of enzyme present is about 1 g in 5 kg of source material. The overall yield for a highly purified protein is likely to be around 5%, and very unlikely to exceed 10%. We might, therefore, hope to obtain around 50 mg of the protein from 5 kg of starting material. Extracting and breaking up this amount of tissue will involve homogenizers and volumes of extract up to 50 L. The extraction often needs to be undertaken in batches and underestimating the levels of starting material is a common problem encountered by the inexperienced when undertaking protein purification studies. It is not surprising that most experimental work in the past has been done on proteins that are present at high levels in the sources, such as seed storage globulins or haemoglobin, or come already in solution as, for example, blood or milk proteins. Isolating proteins for this application is a specialist function, and takes much time and effort. At least 12 months should be allowed for this part of the operation. The ability of modern techniques to facilitate protein purification, obtain amino acid sequence data and isolate the related gene sequence has revolutionized the entire research field. Some of the most important innovations have been made in the ability to sequence entire genomes at comparatively low cost in a matter of days or weeks and next generation sequencing technology has been the subject of many recent reviews, such as, for example, Lusser et al (2012) and Mardis (2013).

1.2.4.2 FINDING THE DNA

The amino acid sequence is used to make probes based on the genetic code. Ideally, at least two probes should be made, aimed at the two extremities of the chain. It is usually possible to identify the N terminal sequence either by directly sequencing the N terminus of the intact protein, or if it is blocked as is often the case, finding the peptide with a blocked N terminus. The C terminus is not so easy, and is a reason for obtaining the complete sequence. Again, if homologous sequences are available they can often help to identify the right sequence. In practice, pairs of probes are often for the N terminus and an internal peptide for this reason. The genetic code is redundant, that is amino acids are specified by more than one codon, with the exception of tryptophan. This means that it is necessary to make a number of nucleotides to be sure of getting the correct one. Since probes need to be made containing at least 12 and preferably 18 nucleotide residues this means that quite a large number may have to be made. This is now relatively easy. The basis of probing is that a complementary string of nucleotides will interact more strongly than non-complementary ones. In practice, the probe is added, and the temperature raised in what is known as *Southern hybridisation*. Non-complementary nucleotides dissociate away at lower temperatures than those with a good match. Radioactive markers or, increasingly, fluorescent markers are used to identify the positive sites.

Once the DNA sequence of a gene is known then the information can be used to express the gene in bacteria or yeast to make the enzyme of interest or the sequence can be transformed into other organisms (e.g. plants) to alter their phenotypes. The development of the *polymerase chain reaction* (PCR) made it possible to amplify sequences directly from the chromosomal DNA. More recently, technological developments have made it possible to synthesize DNA fragments directly for insertion into bacterial cloning vectors.

Bacterial cloning vectors are usually circular DNA structures called *plasmids* that can replicate independently and a wide variety of these are available for purchase from biotechnology suppliers. They often contain regulatory sequences allowing the timing and tissue-specific expression of the transgene. That is to say, they ensure that the chosen sequence is read and gives rise to mRNA with great frequency. The cell produces the mRNA at a rate far above its usual requirements. This is of course exactly what is needed for a method of production of the protein in question. In the case of yeasts, which have become much-favoured production organisms, a promoter that regulates the production of glyceraldehyde 3-phosphate dehydrogenase is used together with a plasmid shuttle vector that can function in both *Escherichia coli* and yeasts. Yeast can in some cases process the pre- and pro-sequences to yield the mature protein, but the exact outcome of cDNA expression is never certain, and yields can be unexpectedly large or small. Another factor is whether the protein is glycosylated, since yeast can sometimes do this, though *E. coli* can never glycosylate. Signals for secretion are also very important. It is far easier to isolate an enzyme if it is secreted into the medium, than if it is necessary to collect and break open the cells. It is thought that pre- and pro-sequences may be involved in secretion, and that in some cases they are removed as the protein passes through the membrane.

In one or two examples, extra chains have been added – such as a string of lysine residues at one end of the sequence – by incorporating the appropriate nucleotides into the cDNA. The strongly positively charged residues make isolation of the protein easy, but they then may have to be removed by treatment with a protease, post purification.

Another difficult problem is that the expressed protein may form so-called inclusion bodies. These are simply lumps of protein which form aggregated and incorrectly folded overproduced enzyme. In some cases, active enzyme can be recovered by dissolving the aggregates in a denaturing solvent such as guanidine hydrochloride or SDS solution, and then using the techniques already established to remove the unfolding agent in such a way as to obtain the correctly folded enzyme. Yields are often poor. It has been claimed that cyclic dextrans (see Chapter 6) are useful in sequestering SDS in this operation, and this could become a significant use for them.

The failure to fold correctly is probably due to the inadequacy of the chaperonin system in the transformed organism. Chaperonins are proteins that are involved in the folding process for other proteins, and for many of them they appear to be essential. It appears likely that the system is simply overwhelmed when heavy overproduction is used, though this is a matter of current research. In one case, the chaperonin cDNA was incorporated into the vector, so that the supply of chaperonin was also increased; this did improve correct folding.

Chaperonins, such as GroES and GroEL, were not known when heterologous expression was first attempted. At the time it was thought that folding would not be a problem, since it was known that a number of small proteins can spontaneously fold *in vitro* without difficulty. This proved to be over-optimistic. Instead it was thought that the next problem – targeting expression to the correct part of multicellular organisms – would be the most difficult to solve. In the event this has proved to be easier than expected. In higher plants, for example, we often wish to target expression to the seeds, but not to other parts of the plant. There are DNA sequences just upstream of the seed globulin DNA which, when incorporated into the vector, seem to bring about just this result. The sequence is highly conserved for the seed storage proteins of legumes.

Vectors in addition usually contain a gene for some component, an enzyme which can easily be detected, or which confers resistance to an antibiotic so as to assist in the selection of transformed organisms. Those that have been transformed are then multiplied in the usual way. In the case of microorganisms, there is usually no great difficulty about this. A low frequency of transformation can easily be made good by the enormous rate of multiplication, and it requires a relatively short time between selecting the transformed organism and proving that it is making the correct protein, and having enough to start production.

1.2.5 COSTS

The costs of developing new products can run into $10 millions and if genetic methods are involved, direct genetic modification, especially related to food products for human consumption, has not been well received by the public due to safety concerns. The focus instead has been on obtaining similar products by conventional breeding using marker-assisted selection and crop-wild species relatives "CWRs", and this has led to some spectacular successes – but can take longer.

In this book, we will now consider in a little more detail the biotechnology of polysaccharides and some of the fundamental properties underpinning it for some of the more important polysaccharides and associated enzymes. For convenience, we have tackled this by considering each of the classes of polysaccharide in turn: *structural polysaccharides* (such as pectins, cellulose, xylan and the mannans locust bean gum, guar and konjac mannan, exudate gums and hyaluronic acid), *storage polysaccharides* (starch, glycogen and the fructans inulin and levan), *marine polysaccharides* (alginate, the principal carrageenans (κ, ι and λ), agar, agarose, chitosan and chitin derivatives), and finally *bacterial and synthetic polysaccharides* (bacterial alginate, xanthan, bacterial dextran, pullulan, scleroglucan or schizophyllan and the cyclodextrins). In each case, we keep structural information to the bare minimum, and try to highlight how these saccharides are useful (and how in some cases they can be made more useful by enzymatic or chemical modification) particularly from a food, pharmaceutical and biomedical standpoint, without excluding their usefulness in other areas such as textiles and printing, cosmetics and the oil industry.

FURTHER READING

Introduction to Polysaccharides

Candy, D.J. (1980) *Biological Functions of Carbohydrates*, Glasgow and London: Blackie & Son.

Habibi, Y. and Lucia, L.A. (2012) *Polysaccharide Building Blocks: A Sustainable Approach to the Development of Renewable Biomaterials*, New York, NY: Wiley-Blackwell.

Ramawat, K.G. and Mérillon, J.M. (eds) (2015) *Polysaccharides: Bioactivity and Biotechnology*, Berlin: Springer-Verlag.

Whistler, R.L. and McMiller, J.B. (eds) (1993) *Industrial Gums. Polysaccharides and Their Derivatives*, 3rd Edition, New York, NY: Academic Press.

Williams, P. (ed) (2011) *Renewable Resources for Functional Polymers and Biomaterials: Polysaccharides, Proteins and Polyesters*, Cambridge: Royal Society of Chemistry.

Polysaccharide structure

Dumitriu, S. (ed). (2004) *Polysaccharides, Structural Diversity and Functional Versatility*, 2nd Edition, New York, NY: Taylor & Francis.

Heinze, T. (2010) *Polysaccharides I: Structure, Characterisation and Use (Advances in Polymer Science)*, Berlin: Springer-Verlag.

Basic Theory of Solution Measurement

Serydujk, I.N., Zaccai, N.R. and Zaccai, J. (2007) *Methods in Molecular Biophysics. Structure, Dynamics and Function*, Cambridge: Cambridge University Press.

Polysaccharide Rheology

Lapasin, R. and Pricl, S. (1995) *Rheology of Industrial Polysaccharides. Theory and Applications*, London: Blackie.

Polysaccharide Identification or Fingerprinting

Baird, J.K. (1993) Analysis of gums in foods, in Whistler, R.L. and McMiller, J.B. (eds) *Industrial Gums. Polysaccharides and Their Derivatives*, 3rd Edition, Chap. 23, New York, NY: Academic Press.

SPECIFIC REFERENCES

References referred to in the text not already included in the Further Reading Section:

Alexander, L.E. (1979) *X-Ray Diffraction Methods in Polymer Science*, Huntington, NY: Krieger.

Atkins, E.D.T., Mackie, W. and Smolko, E.E. (1970) Crystalline structures of alginic acids, *Nature*, 225, 626–628.

Burchard, W. (1992) Static and dynamic light scattering approaches to structure determination of biopolymers, in Harding, S.E., Sattelle, D.B. and Bloomfield, V.A. (eds) *Laser Light Scattering in Biochemistry*, Chap. 1, p. 291, Figure 1, Cambridge: Royal Society of Chemistry.

Clark, A.H. (1994) X-ray scattering and diffraction, in Ross-Murphy, S.B. (ed) *Physical Techniques for the Study of Food Biopolymers*, Chap. 3, London: Blackie.

Deacon, M.P., McGurk, S., Roberts, C.J., Williams, P.M., Tendler, S.B., Davies, M.C., Davis, S.S. and Harding, S.E. (2000) Atomic force microscopy of gastric mucin and chitosan mucoadhesive systems, *Biochemical Journal*, 348, 557–563.

Dhami, R., Harding, S.E., Jones, T., Hughes, T., Mitchell, J.R. and To, K.-M. (1995) Physico-chemical studies on a commercial food grade xanthan I. Characterisation by sedimentation velocity, sedimentation equilibrium and viscometry, *Carbohydrate Polymers*, 27, 93–99.

Gamini, X., Paoletti, S. and Zanetti, F. (1992) Chain rigidity of polyuronates: Static light scattering of aqueous solutions of hyaluronate and alginate, in Harding, S.E., Sattelle, D.B. and Bloomfield, V.A. (eds) *Laser Light Scattering in Biochemistry*, Chap. 20, Cambridge: Royal Society of Chemistry.

Grálen, N. (1944) Sedimentation and diffusion measurements on cellulose and cellulose derivatives, PhD dissertation, University of Uppsala, Sweden.

Grasdalen, H. (1983) High-field, ^{1}H-n.m.r. spectroscopy of alginate: Sequential structure and linkage conformations, *Carbohydrate Research*, 118, 255–260.

Harding, S.E. (1997) The intrinsic viscosity of biological macromolecules. Progress in measurement, interpretation and application to structure in dilute solution, *Progress in Biophysics and Molecular Biology*, 68, 207–262.

Harding, S.E. (2013) Intrinsic viscosity, in Roberts, G.C.K. (ed) *Encyclopedia of Biophysics*, pp. 1123–1129, Berlin: Springer Verlag.

Harding, S.E., Abdelhameed, A.S., Morris, G.A., Adams, G.G., Laloux, O., Cerny, L, Bonnier, B., Duvivier, P., Conrath, K. and L'Enfant, C. (2012) Solution properties of capsular polysaccharides from Streptococcus pneumoniae, *Carbohydrate Polymers*, 90, 237–242.

Harding, S.E., Adams, G.G., Almutairi, F.M., Alzahrani, Q., Erten, T., Kok, M.S. and Gillis, R.B. (2015) Ultracentrifuge methods for the analysis of polysaccharides, glycoconjugates and lignins, *Methods in Enzymology*, 562, 392–349.

Harding, S.E., Adams, G.G. and Gillis, R.B. (2016) Molecular weight analysis of starches: Which technique? *Starch/Stärke*. 68, 1–8.

Harding, S.E., Hill, S.E. and Mitchell, J.R. (eds) (1995) *Biopolymer Mixtures*, Nottingham: Nottingham University Press.

Harding, S.E., Schuck, P., Abdelhameed, A.S., Adams, G.G., Kök, M.S. and Morris, G.A. (2011) Extended Fujita approach to the molecular weight distribution of polysaccharides and other polymeric systems *Methods*, 54, 136–144.

Heinze, T., Nikolajski, M., Daus, S., Besong, T.M.D., Michaelis, N., Berlin, P., Morris, G.A., Rowe, A.J. and Harding, S.E. (2011) Protein-like oligomerisation of carbohydrates, *Angewandte Chemie International Edition*, 50, 8602–8604.

Hermansson, A.M. and Langton, M. (1994) Electron microscopy, in Ross-Murphy, S.B. (ed) *Physical Techniques for the Study of Food Biopolymers*, Chap. 6, Glasgow: Blackie & Son.

Horton, J.C., Harding, S.E. and Mitchell, J.R. (1991) Gel permeation chromatography—Multi angle laser light scattering characterization of the molecular mass distribution of "Pronova" sodium alginate, *Biochemical Society Transactions*, 19, 510–511.

Jumel, K., Fiebrig, I. and Harding, S.E. (1996) Rapid size distribution and purity analysis of gastric mucus glycoproteins by size exclusion chromatography/multi-angle laser light scattering, *International Journal of Biological Macromolecules*, 18, 133–139.

Jung, A. and Berlin, P. (2005) New water-soluble and film-forming aminocellulose tosylates as enzyme support matrices with Cu^{2+}-chelating properties, *Cellulose*, 12, 67–84.

Launay, B., Doublier, J.L. and Cuvelier, G. (1986) Flow properties of aqueous solutions and dispersions of polysaccharides, in Mitchell, J.R. and Ledward, D.A. (eds) *Functional Properties of Food Macromolecules*, Chap. 1, London: Elsevier Applied Science.

Lusser, M., Parisi, C., Plan, D. and Rodriguez-Cerezo, E. (2012) Deployment of new biotechnologies in plant breeding, *Nature Biotechnology*, 30, 231–239.

Mardis, E.R. (2013) Next-generation sequencing platforms, *Annual Review of Analytical Chemistry*, 6, 287–303.

Nikolajski, M., Adams, G.G., Gillis, R.B., Besong, D.T., Rowe, A.J., Heinze, T. and Harding, S.E. (2014) Protein-like fully reversible tetramerisation and super-association of an aminocellulose, *Nature Scientific Reports*, 4, 3861.

Pavlov, G.M., Rowe, A.J. and Harding, S.E. (1997) Conformation zoning of large molecules using the analytical ultracentrifuge, *Trends in Analytical Chemistry*, 16, 401–405.

Rollings, J.E. (1992) Use of on-line laser light scattering coupled to chromatographic separations for the determination of molecular weight, branching, size and shape distributions of polysaccharides, in Harding, S.E., Sattelle, D.B. and Bloomfield, V.A. (eds) *Laser Light Scattering in Biochemistry*, Chap. 19, Cambridge: Royal Society of Chemistry.

Satyajit, S., Vidyarthi, A.S. and Prasad, D. (2014) RNA interference: Concept to reality in crop improvement, *Planta*, 239, 543–564.

Schuck, P., Gillis, R.B., Besong, D., Almutairi, F.M., Adams, G.G., Rowe, A.J. and Harding, S.E. (2014) SEDFIT-MSTAR: Molecular weight and molecular weight distribution analysis of polymers by sedimentation equilibrium in the ultracentrifuge. *Analyst*. 139, 79–92.

Smidsrød, O. and Andresen, I.L. (1979) *Biopolymerkjemi*, Trondheim, Norway: Tapir Press.

Smidsrød, O. and Haug, A. (1971) Estimation of the relative stiffness of the molecular chain in polyelectrolytes from measurements of viscosity at different ionic strengths, *Biopolymers*, 10, 1213–1227.

Stokke, B.T. and Elgsaeter, A. (1991) Electron microscopy of carbohydrate polymers, *Advance Carbohydrate Analysis*, 1, 195–247.

Taylor, R.J., Welsh, E.J. and Morris, E.R. (1983) *Carbohydrates*, Advanced Educational Booklet of 1982–3, Figure 19, London: Unilever.

Wyatt, P.J. (1992) Combined differential light scattering with various liquid chromatography separation techniques, in Harding, S.E., Sattelle, D.B. and Bloomfield, V.A. (eds) *Laser Light Scattering in Biochemistry*, Chap. 3, Cambridge: Royal Society of Chemistry.

Wyatt, P.J. (2013) Multiangle light scattering from separated samples (MALS with SEC or FFF), in Roberts, G.C.K. (ed) *Encyclopedia of Biophysics*, pp. 1618–1637, Berlin: Springer Verlag.

Structural Polysaccharides

This chapter considers a group of important polysaccharides which *in vivo* are involved with the structure of land-based plants or parts of plants: fruits (pectins), general cell walls (cellulose), wood (xylans) and gum plants (the galactomannans locust bean gum and guar, and the glucomannan konjac mannan). Three exudate gums (involved in trees and shrubs with structural repair) are also briefly considered: gum arabic, tragacanth gum and karaya gum, particularly because of their useful adhesive properties. Structural polysaccharides from marine-based organisms or microorganisms are considered in Chapters 5 and 6, respectively, but we finish this chapter with a consideration of glycosaminoglycans involved with the structure of animal tissue, focussing on what is probably the biotechnologically most important, namely hyaluronic acid.

2.1 PECTIN

Pectins are polymers of galacturonic acid linked by $\alpha(1{\rightarrow}4)$ linked bonds:

$$...{\rightarrow}4)\ \alpha\text{-}\mathrm{D}\text{-}GalpA\text{-}(1{\rightarrow}...$$

where the carboxyl groups are methylated to varying degree. The linear galacturonan chain also contains a number of rhamnose (L-Rha*p*) residues which through $\alpha(1{\rightarrow}2)$ links impose a bend in the chain, and may also form a branch point through which arabinogalactans are linked in the cell wall structure. Some pectins contain a high proportion of these (intrachain) rhamnose residues, and one has been described with alternating galacturonic and rhamnose residues (Bacic et al., 1988). Some of the structures that have been found are shown in Figure 2.1 and include as side chains fucose L-Fuc*p* and xylose D-Xyl*p* residues as well as the predominant arabinans and galactans in the side chains. The branch points do not always derive from the rhamnopyranose structure in the main chain and not all rhamnose residues form branch points. Some hydroxyl groups may be acetylated on the galacturonic acid residues. The arabinogalactan side chains may also be branched. Regions with extensive branching are popularly referred to as 'hairy regions' and these, together with the degree of methylation (i.e. esterification) of acidic residues, are

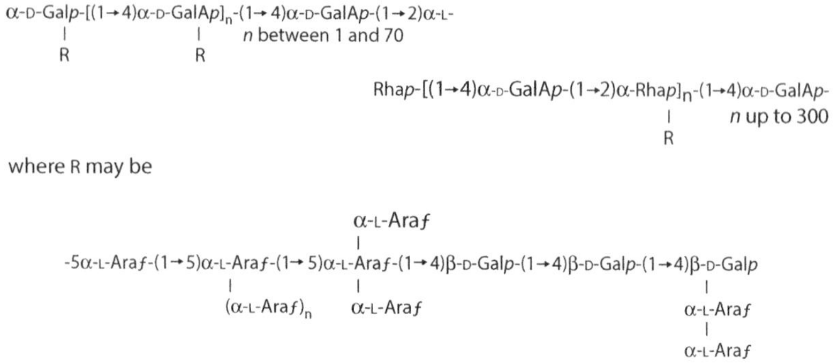

α-D-Gal*p*-[(1→4)α-D-GalA*p*]ₙ-(1→4)α-D-GalA*p*-(1→2)α-L-
 | | *n* between 1 and 70
 R   R

                                    Rha*p*-[(1→4)α-D-GalA*p*-(1→2)α-Rha*p*]ₙ-(1→4)α-D-GalA*p*-
                                                                   | *n* up to 300
                                                                   R

where R may be

                                    α-L-Ara*f*
                                       |
-5α-L-Ara*f*-(1→5)α-L-Ara*f*-(1→5)α-L-Ara*f*-(1→4)β-D-Gal*p*-(1→4)β-D-Gal*p*-(1→4)β-D-Gal*p*
                  | | |
             (α-L-Ara*f*)ₙ   α-L-Ara*f*   α-L-Ara*f*
                                                                               |
                                                                          α-L-Ara*f*

FIGURE 2.1   Some typical pectin compositions and sequences.

TABLE 2.1   Fundamental Properties of Pectins

Property

Polymer type	Polyanionic: interrupted repeat
Conformation type	Zone B/C: rod/semi-flexible coil
Solubility	Depends on degree of esterification (DE) and temperature
Molecular weight	~200 kDa
Intrinsic viscosity [η]	~600 mL g^{-1}
Sedimentation coefficient, $s_{20,w}$	~(1.9 ± 0.2) S
Wales–van Holde ratio, $k_s/[\eta]$	~0.2
Mark–Houwink a and b coefficients	~1, 0.2
Gelation properties	Strong gels particularly in presence of divalent metal ions (for low DE pectin) or small sugars at low pH (high DE pectin)
Key sites for biomodification	COO$^-$: sites of esterification

considered to have a strong influence on their physical properties. These properties, based on results for a citrus pectin, are summarized in Table 2.1.

2.1.1 OCCURRENCE

Pectins occur as a major component (up to 35%) in the cell walls of terrestrial plants, while the algal polysaccharides like carrageenans or alginates seem to serve a similar function in the cell walls of marine algae. Plant cells have a lamellar structure and consist of layers of cellulose fibres, often oriented at right angles to each other, separated by layers of pectin. In addition, woody plants and cereals contain lignin and xylans. Lignin, whilst not a polysaccharide (being a polyphenyl propanoid) nevertheless deserves some consideration since its presence has considerable effects on the extraction and hence the utilization of the other cell wall components.

Both pectins and xylans have branched chain structures and are charged molecules at the physiological pH. They undoubtedly contribute, along with the oriented fibrils of cellulose, to the mechanical properties of the cell wall. The fact that they are charged must also have a major effect on the penetration of the cell wall by charged smaller molecules.

In practice, the charges appear to be neutralized by calcium ions *in vivo* to such an extent that calcium can be regarded as part of the pectin in its normal physiological state. In some circumstances, however, the calcium may be replaced by other divalent ions; this may be the mechanism by which algae are sometimes found to have selectively concentrated ions such as rubidium in the vicinity of radioactive discharges.

The occurrence of pectins in cell walls may be determined in different ways, with the heterogeneity in the cell walls described by Burton et al. (2010) and the polysaccharide composition of the cell walls described by Pettolini et al. (2012) and Sousa et al. (2015).

2.1.2 EXTRACTION, PRODUCTION AND USES OF PECTINS

There are broadly two ways in which a biotechnology approach might improve methods for the pectin extraction, or indeed any other extraction process. They are first of all to use enzymes to break down the biological structures involved, thus making extraction faster or possibly more complete. There are surprisingly few examples where this has been explicitly done. The other way is to modify the structure of the source material by plant breeding or more recently by *marker assisted selection* (Section 2.1.7). Pectins provide one of the best examples of where a degradation process has been modified by gene manipulation to improve properties of commercial importance.

'High DE' (degree of esterification) or 'high methoxy' (HM) pectin production is typically 12,000 tonnes p.a. (per annum) and 'low DE' (LM) about 5000 tonnes. For comparison, carrageenans at about 18,000 tonnes p.a. and alginates at about 6000 tonnes p.a. are to some extent competitors.

Pectins are made either from apple pulp, a by-product of cider manufacture, or from the peel of citrus fruits such as limes, the preferred source, lemons or oranges, by-products of fruit juice manufacture. There is no other significant source at present though sugar beet residues might become one and, unlike the situation for many other polysaccharides, no crop is grown solely as a pectin source (May, 1990). The extraction process is based on the use of raw materials that have been dried for storage, and are usually processed on the spot. In some cases the north European factories, based on the use of apple pomace, a very seasonal material, have imported citrus peels so as to maintain year-round operation. Flammability of dried peels is a problem and spontaneous combustion is not unknown.

Extraction, after comminution of the peel, is with hot dilute nitric acid. There are problems associated with the viscosity of the extract and the degree of breakdown of the solids which can lead to difficulty in subsequent clarification. There is a close relationship between the temperature–time characteristic and the molecular weight and degree of methylation of the final product. Figure 2.2 shows some of the conditions for extraction from apple residue – other raw materials will be different. The molecular weight (weight average) of pectins varies between 30,000 and 300,000 Da depending on the source and isolation method;

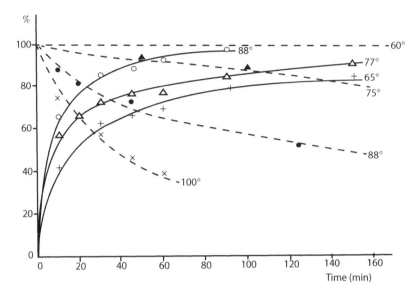

FIGURE 2.2 The yield and molecular weights of pectins isolated at different temperatures. − − − − indicates the relative molecular weight; ──── indicate the relative yield. The greater the yield, the lower the molecular weight. (Bergenstahl [1995]. Reproduced by permission of the Royal Society of Chemistry.)

although the low pH prevents most enzyme activity, there may be some contribution to breakdown in the early stages of the process. Since pectins are probably covalently linked to the other components of the cell wall, some kind of acid hydrolysis is necessary to extract them and the yields are surprisingly high.

Around 60% of the cell wall pectin is recovered. Endogalacturonases have been used to improve recovery but only in experimental work. It is possible that if efficient cell wall–breaking enzymes that do not at the same time hydrolyse pectins should ever be developed, they might be worth using to improve yields. Since this is one of those situations, common in by-product processing, where the raw material is cheap and plentiful, the only economic justification would be the more efficient utilization of plant, and the cost of the enzymes will have to be minimal.

The residues left after the pectin extraction have no known use and represent a disposal problem. They tend to be given, at little or no cost, to anyone who will remove them. They can be added in small amounts to cattle feed, used as soil conditioners and used even as fuel in some countries. This is also the fate of surplus citrus peel not required for pectin production. A very small tonnage of orange peel is used in making a Belgian beer, and the oil can be used in chocolate making.

A possible new source of pectins is the residue left after sugar production from beet. The pectin is of relatively low molecular weight, but it is not clear whether this is due to the extraction process or is inherent, and is also fairly highly acetylated. This means that its properties are different from those of the conventional commercial pectins, and the products containing them would have to be reformulated. On the other hand, it might well be considered for new products which might reveal advantages in comparison with the older materials.

In terms of the by-product status it is similar to apple pomace, available in large amounts, and an embarrassment to those who must dispose of it. It must be emphasized that this kind of residue is becoming increasingly difficult to dispose of in Europe, and a major theme of biotechnology and commercial development is generally to find ways of increasing the value by finding uses for such residues. The biological oxygen demand is very high, and the old solution of tipping into rivers is no longer acceptable. Landfill disposal is expensive and of marginal acceptability, and the position has now been reached where some processes may not be operated (at least in Europe) because the residues are too difficult to get rid of. There are many other potential pectin sources, but whether or not they are used will depend on local circumstances except in the unlikely event that they are found to have unusual and valuable properties in use.

After extraction at low pH which ensures that the carboxyl groups are un-ionized and thus do not bind calcium, the extract is clarified by filtration using kieselguhr, a diatomaceous earth especially useful for difficult filtrations. In some processes amylase may be used to break down starch. The extract is then concentrated by evaporation and may be held for a few days to allow hydrolysis to continue if a low methoxy-pectin is intended. Otherwise it is mixed with ethanol to 'precipitate' the pectin. This actually separates as gelatinous particles which are allowed to undergo syneresis to assist in settling, followed by washing with aqueous alcohols and pH adjustment. The quality of the pectin produced is dependent on empirical knowledge of temperature, time and pH relationships built up over many years, concentrated on particular raw material supplies. An outline of a typical large-scale production process is shown in Figure 2.3.

An alternative process uses adsorption on colloidal aluminium hydroxide followed by immediate pH adjustment and although it has been used for citrus raw materials it is interfered with by citrate, which must be removed with care. It is a process which can be used to avoid prolonged low pH exposure and thus obtain HM-pectins. Washing and removal of the aluminium make use of the insolubility of pectins in acid–ethanol–water mixtures.

These processes are able to produce the whole range of methoxy-pectins currently in use, down to those with virtually no methyl esters remaining. This derivative, sometimes referred to as pectate, is neutralized and available as the sodium or ammonium salt.

Amidated pectins are also produced. If pectin is suspended in ethanol and treated with ammonia, a heterogeneous reaction leads to the replacement of methyl groups by amides, and these derivatives have found applications.

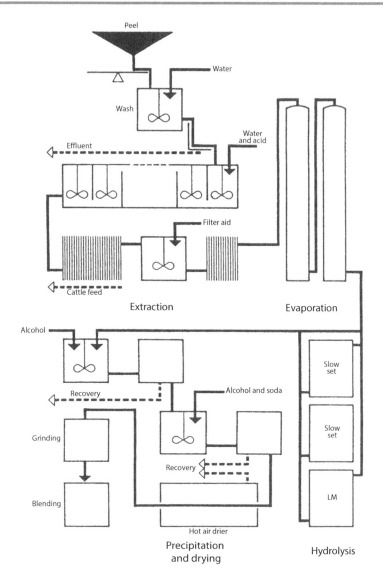

FIGURE 2.3   An outline of a large-scale process for the manufacture of pectins from apples. (Bergenstahl [1995]. Reproduced by permission of the Royal Society of Chemistry.)

2.1.3 ENZYMATIC MODIFICATION

There are a number of enzymes known to act on pectins. Amongst these, *pectate lyase* cleaves bonds between methylgalacturonate residues. Pectate lyases carry out a similar reaction between residues that lack the methyl group, to yield an unsaturated uronide. Figure 2.4 illustrates this reaction. There are both exo- and endo-enzymes and the activity appears to be widespread in plants and microorganisms. They are usually calcium-requiring and have fairly high pH optima. They are significantly involved in fruit ripening (see Section 2.1.7).

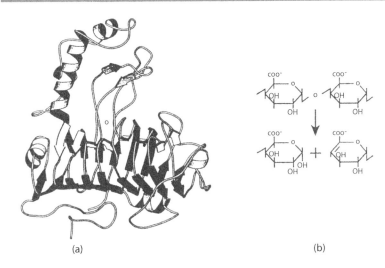

(a)   (b)

FIGURE 2.4   A structure for pectate lyase. The flat arrows indicate β-sheets, three of which are the main structural element. There are also two helices. The small circle shows the position of the active site calcium ion. To the right is shown the reaction catalyzed by the enzyme. (Reprinted from Pickersgill, R. et al., in Geisow, M.J. and Epton, R. (eds), *Perspectives on Protein Engineering*, Mayflower Worldwide Ltd., Birmingham, 1995.)

Pectate hydrolases act on both esterified and carboxyl-rich pectins, and break the α(1→4) glycosidic bonds. The exo-enzyme produces galacturonic acid from the end of the chain, but endo-varieties are also known which can act at any point in the chain. They do not act randomly, as is shown in Figure 2.5. This compares the effect of random fission with that actually determined for three industrial pectin samples. In every case, the production of smaller fragments was greater than predicted, suggesting a tendency to prefer residues near the ends of the chains. These are much more significant since they can have a large effect on the molecular weight during isolation. They are also significant during fruit ripening. There are probably enzymes capable of removing rhamnose residues as well, although these have not been investigated. Another important enzyme is *pectin esterase*, which removes the methyl groups. The gelation properties of pectin are dependent on the level of esterification and production processes are manipulated to provide a variety of pectins with different DE. So far this has been done by the controlled alkali hydrolysis, but this has disadvantages. The same conditions always cause some fission of the main chain with a concomitant reduction in molecular weight which is undesirable. In addition de-esterification is random and never complete.

Enzymatic hydrolysis removes methyl groups from one end of the chain progressively and completely and can be controlled. Thus, it offers a range of products which will actually be different from the random alkali hydrolysis process. It had the disadvantage that sources of the esterase tended to be contaminated by pectin hydrolases so that the molecular weight was in danger of reduction. This difficulty has been

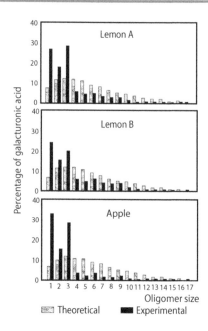

FIGURE 2.5 The effect of an endopolygalacturonidase isolated from *Kluyveromyces fragilis* on three pectins with high-methoxy content. The grey bars show the predicted results of the size of the fission products if the methoxy groups are randomly distributed. They are not as is shown by the actual results, and also differ between samples. (Kravtchenko, T.P., Parker, A. and Trespoey, A. (1995), *Food Macromolecules and Colloids,* Royal Society of Chemistry, Cambridge. Reproduced by permission of the Royal Society of Chemistry.)

resolved (Cheetham, 1995). Commercial papain, made from papaya latex and of course used as a protease, contains a pectin esterase activity, but no pectin hydrolase activity, and thus offers a promising source of the requisite enzyme. This is much more attractive than the alternative which would have been to obtain the gene for the esterase and express it in some convenient organism. This is probably feasible and may be done, but it does not look attractive compared with an easy conventional source.

Finally, the role of the side chains is not fully understood in functional terms, but enzymes probably exist capable of removing or modifying them. Yet there is no case for doing this.

2.1.4 ENZYMES IN FRUIT JUICES AND PROCESS OPERATIONS

Although pectin is made from the residues after extraction of juice, some pectin is extracted into the juice, and has a large effect on the properties. Methods have been developed using pectinases to assist in both the extraction process and control of juice properties (Lea, 1995). These have found most application in cider making and citrus fruit juice manufacture.

When apple juice is expressed it becomes brown and turbid, due to oxidation of phenolics and then complex formation with the acidic pectins. The latter are increased by the action of endogenous pectin esterase. This turbidity can persist through to the final fermented product and be very difficult to clarify. In traditional processes, the juice was held until so much pectate had been produced by the action of the esterase that it formed a distinct gel phase together with free calcium present in the juice. It should be noted that apples do not have a polygalacturonidase, so there is no degradation of the main chain. Particulate matter was trapped in the gel and could thus be removed, because this also reduced the free nitrogen, subsequent yeast growth was restricted. Of course additional calcium and yeast nutrients can be added. The process is still being used on a small scale by some traditional brewers and suffers from unreliable levels of pectin esterase in the apples used.

Large-scale manufacture of apple juice has employed added pectin and polygalacturonase for the last 60 years, the enzymes being obtained from fungi. The mechanism by which clarification occurs is complex, and probably involves changes in the surface charge of suspended particles, as a result of enzyme action, followed by electrostatic aggregation and sedimentation.

In a combined process, enzymes which degrade pectin are sometimes added to the apple pulp, where they improve not only the yield but also the mechanical properties thus making juice expression easier. A general disadvantage of the use of enzymes in this operation is that it is necessary to allow time for the enzymes to act, which may cause other problems.

A novel problem found with some enzyme-treated pulps was the appearance of haze, apparently due to the arabinan side chains left over from hydrolysis of the main chain. Arabinases have been made available to deal with this problem. In a few cases, amylases are used to remove suspended starch particles, although these are not usually present in apple juice.

In a fairly obvious extension of pectinase treatment, attempts have been made to combine it with cellulases to achieve what amounts to total liquefaction of the fruit. These are still experimental, and may have problems of consumer acceptance.

Although great efforts are made to obtain clarity in apple juice, the opposite is wanted for orange juice, where the suspended solids carry most of the flavour. The shelf life and general properties of the suspension depend on the level and viscosity properties of the pectates present. There is sometimes a need to reduce the viscosity, while avoiding charge effects due to methyl esterase activity. Thus, enzyme preparations containing only galacturonidase activity and no esterase activity are needed and have been made available. They are added after heat treatments to destroy the endogenous enzymes, which do include an esterase. A more ambitious approach seeks to break the cell walls in such a way that the individual cells are retained but float free. This is a technique developed for plant tissue culture and is linked to ideas about culturing the flavour-containing cells of valuable plants such as strawberries. It has always been found to be economically non-viable in the past.

Finally, we should point out that the activity of pectin esterase produces methanol which is toxic, and while processes to make isolated pectins have plenty of opportunity for it to evaporate, fruit juices usually contain around 50 ppm (parts per million) which is not considered to be a problem. It may be increased 10-fold by the use of enzyme treatments and some sort of monitoring should be considered although no regulatory difficulties have arisen as yet.

Other applications of pectinases have been reported. In the depulping of coffee beans, for example, while traditional processes rely on indigenous enzymes, addition of pectinases can give better results. The real gain is in faster operation, minimizing the risk of microbial contamination, always a serious problem in tropical crops.

2.1.5 BIOSYNTHESIS

The substrate for pectin synthesis is UDP-GalA and a similar activated rhamnose is presumably used to insert the rhamnose residues. Methylation occurs during polymerization with S-adenosylmethionine as the methyl donor and probably acetyl CoA where residues are acetylated. The mechanism does not appear promising for manipulation though gene deletion through the use of adventitious mutants may provide some opportunity for variation. Lack of the enzymes responsible for rhamnose incorporation could lead to homopolymer galacturonan, though the cell walls would certainly be defective. Petite mutants of yeast which lack cell walls are apparently unable to make cellulose. The enzymes involved have not been isolated or characterized, but as indicated elsewhere are membrane bound and are likely to be difficult.

2.1.6 DEGRADATION *IN VIVO*

Control of the degradation of pectins is one of the major commercial applications of polysaccharide biotechnology so far. Human consumption of fruit is quite incidental to its true function which is to facilitate the distribution of the seeds. As a result, every fruit has a ripening mechanism which includes degradation of the structure so as to lead to the release of the seeds. In some cases this is the result largely of microbial attack, but more often there are also inbuilt mechanisms which lead to a whole variety of changes amongst which is the softening and dissolution of the cell walls.

There is considerable wastage between the picking of fruit and its end use, whether this is consumption as fruit or processing to make, for example, tomato sauce. There would be more than enough economic advantage if the ripening of fruit could be controlled, and in the late 1980s this became the focus of *genetic modification* (GM) technologies. Soon after these methodologies became controversial in Europe following questions raised in the media about the safety of such technologies. In response, the focus has been on obtaining the same goals as GM but by conventional breeding using *marker assisted selection* and *crop-wild species relatives.*

2.1.7. MARKER ASSISTED SELECTION: THE TOMATO RIPENING STORY

As an example of marker assisted selection, in 2016 in an important breakthrough made by an interdisciplinary team of scientists led by Graham Seymour, Professor of Plant Biotechnology at the University of Nottingham, a gene encoding a key enzyme responsible for tomato softening during ripening – a *pectate lyase* (see Figure 2.4) – was identified. Commercial plant breeders had been working continuously trying to find cultivars which where high yielding, more nutritious and better tasting – but some of the best tasting ones had short shelf lives.

The gene the 'TOMNET' team identified and sequenced was a pectate lyase, an enzyme which degrades the cell wall pectin polysaccharides. By switching the gene off the tomatoes soften much more slowly, and this correlates with a reduction in degradation of the molecular weights (molar mass) of the pectins (Figure 2.6), as determined by sedimentation equilibrium in the analytical ultracentrifuge (Chapter 1). Writing in the popular magazine *Science Daily* Prof Seymour said "In laboratory experiments we have demonstrated that if this gene is turned off, the fruit soften much more slowly, but still show normal changes in colour and the accumulation of taste compounds such as acids, sugars and aroma volatiles. Natural varation exists in the levels of pectate lyase gene expression in wild relatives of cultivated tomato and these can be used for conventional breeding purposes. This discovery can provide a means to refine the control of fruit softening in modern tomato cultivars".

The discovery, published in *Nature Biotechnology* is considered as forming the basis for the generation of new varieties of better tasting tomatoes with improved postharvest life through conventional plant breeding.

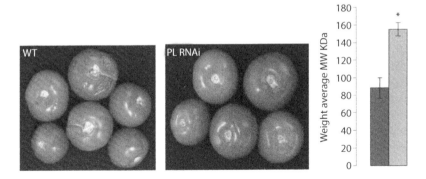

FIGURE 2.6. Silencing pectate lyase enhances shelf life. Wild-type (WT) Ailsa craig (left) and pectate lyase (PL) inhibited by RNA interference "RNAi" (centre). Both photographs show tomatoes 7 days and 14 days after ripening. For the PL inhibited, no difference is discernible. Right: Weight average molecular weights (from analytical ultracentrifugation) for WT pectin (dark gray) and RNAi pectin (light gray) 7 days after ripening: the latter is close to the value for the unripe fruit: this lack of degradation is borne out by the photographs. (Adapted from Uluisik et al., *Nature Biotechnology*, 2016. With permission.)

2.1.8 UTILIZATION AS A GEL AGENT

Pectins have long been recognized as the main gel-forming agents in jams and fruit-based preserves. It is possible to buy sucrose with pectin added specially for jam making, and manufacturers add isolated pectins to jams to ensure reproducible and controlled gel formation. Anyone who has attempted to make strawberry jam will know that it is particularly difficult, and pectins were first commercially used for this fruit. Jam making is still probably the main use.

Pectin can form two quite different kinds of gel, and within these there is considerable variation in the physical properties of the gels, such as strength and rate or temperature of setting. The first kind of gel is *thermolabile*. These gels form when a solution of pectin is cooled and melt again on heating. It is important that the pH is sufficiently low that the carboxyl groups are *not* ionized but are instead available to form hydrogen bonds as well as being uncharged. Charged groups cause repulsions between chains that tend to prevent interchain links from forming. Hydrogen bonds are the main interchain links that lead to gelation. It has been reported that urea, a hydrogen bond breaker, breaks interchain links in this type of pectin gel. This means that the pH should be below 4, since most carboxyls have a pK (the pK is the pH at which the group is 50% ionized) around 4.5. The formation of interchain links will also depend on the availability of carboxyl groups and thus on the degree of methylation. This is why manufacturers make and sell pectins with different DE values. The lower the DE the greater the frequency of interchain charge repulsions which leads to gels of reduced mechanical strength, forming more slowly and stable at lower temperatures. Broadly the dividing line is around 50% DE. Pectins with DE 70%–80% are the so-called *rapid set* while DE 55%–60% are the *slow set*. Another factor which is certainly important is the average molecular weight – a low molecular weight is associated with lower mechanical strength. It is not immediately obvious why this should be so, and requires detailed consideration of gel structures to understand. Arabinogalactan side chains must also be involved, though the degree to which they may influence interchain interactions, in either direction, is unknown, as is the effect of rhamnose residues in the main chain itself. It is believed that links are formed between long unbranched segments of the main chain, a very common feature of polysaccharide gelation mechanisms, and there is evidence that greater branching correlates with weaker gels. There is one report that interaction between arabinogalactan side chains may lead to gelation in apricot juice.

The second gelation mechanism shown by pectins involves calcium as a divalent cation, where addition of calcium salts brings about rapid gel formation, the gels being *thermostable*. This undoubtedly involves the free carboxyl groups and is therefore much greater for low DE pectins. It depends on the availability of runs of polygalacturonic acid, and involves a specific structure sometimes known as the 'egg box'. Similar structures form with other carboxylated polysaccharides such as alginates and carrageenans (Chapter 5) and mixed gels are possible. The relationship

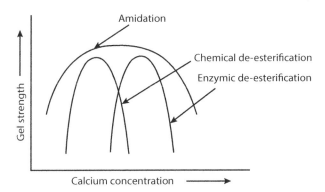

FIGURE 2.7 A diagrammatic representation of the relationship between gel strength and the processing history of the pectin. (Reprinted from New and modified polysaccharides, Vol. 1, Chap. 5, Morris, V.J., *Food Biotechnology*, 1987, with permission from Elsevier.)

between calcium concentration and gel strength depends not only on the DE value and the molecular weight, but also on the way in which de-esterification has been achieved (Figure 2.7). Enzyme de-esterified pectin has not been widely used because it may require approval as a food additive, and the alternative is widely available and apparently meets all identified needs. Pectin esterase from *Aspergillus niger* is a permitted enzyme in the United Kingdom.

In general application both kinds of interchain link will form. It would be surprising to find foods containing no calcium, though it must be free to interact, and in particular dairy products contain high levels of calcium. The amidated pectins are less sensitive to calcium level (Figure 2.7) and are potentially useful for products with marginal or variable calcium levels.

The very rapid formation of calcium induced gels has been used experimentally in several ways. Extrusion of pectate suspensions of fish protein into calcium setting baths can be used to make structured reformed fish products, while a slightly different form of extrusion of mixed blackcurrant juice and pectate into calcium chloride setting baths can yield imitation blackcurrants. Pectate gels have been important in developing the whole field of reformed fruit.

Commercial fruit products, like apple pies, depend on pieces of fruit but inevitably cutting these leaves unusable small fragments which have very little value. It is worthwhile reforming these into larger pieces with gel systems like pectate, where textures close to those of the fruit can be obtained. Note, however, that the texture is that of apple in apple pie, not raw apple which might not be so easy. Yoghurt, which is commercially dominated by its fruit content, can also benefit from this technology.

2.1.9 UTILIZATION AS A COLLOID STABILIZER

It has been known for a long time that polysaccharides can stabilize colloidal sols and suspensions. Simple addition of water-soluble pectate, for example, usually the sodium salt will prevent aggregation and precipitation. They can also stabilize emulsions, for example, of lipid in water or the reverse. In ice cream they can affect the stability of the ice crystal, lipid, water and air mixture. Gold sols, and metal sols in general, have been stabilized with gums for many years.

Despite this there is no very clear idea of the mechanisms involved. It is believed that the polysaccharide is adsorbed to the surface of the particles, where if it is charged it might increase charge–charge repulsions between them, thus slowing down the rate of aggregation. There are instances known where if, for example, the particle already carries a positive charge, adding a negatively charged polysaccharide actually reduces the net surface charge, and leads to faster aggregation. In other instances, the effect is thought to be due to the presence of the polysaccharide chain simply blocking collisions of sufficient energy to lead to cohesion. In slightly different terminology a higher viscosity leads to less interaction, but also slower sedimentation.

A good example of the effect of pectin is shown in Figure 2.8 where an acid milk drink of the *kefir* type (rather like a diluted yoghurt) has its particle size distribution altered by the addition of pectin. Note that this was a commercial HM-pectin at a comparatively high concentration of 0.25% (2.5 g L⁻¹) and that highly charged low methoxy-pectins do not give the same kind of result. Although at first sight, charge repulsions look to be a good candidate as an explanation of the way in which coalescence of the casein particles is prevented, in fact measurements of surface charge show that the forces are too small. The best explanation is that the pectin adsorbs to the casein surface and prevents contact by steric hindrance. Some evidence for this was obtained by sedimenting the particles under fairly high gravitational field and estimating the average volume of the particles: it was increased in the presence of pectin.

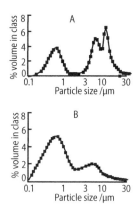

FIGURE 2.8 The particle size distribution in an acid milk drink in the absence (A) and presence (B) of added pectin at the level of 0.25%. The presence of numerous larger particles in the absence of pectin indicates the onset of casein aggregation. (Kravtchenko, T.P. et al., *Food Macromolecules and Colloids*, 1995, by permission of Oxford University Press.)

2.1.10 UTILIZATION IN EMULSIONS

In the case of emulsions, the effect can be rationalized by assuming adsorption to the interface and changes in surface energy as a result. Whatever the mechanisms the results are useful in food formulations; pectins and other polysaccharides are widely used. The uses tend to be at very low levels. Colloid stabilization is seen at the 'ppm' concentration level as would be expected for what are essentially surface adsorption effects.

Significantly, a potentially major use has been developed which involves the gels obtained by mixing gelatin and pectin into lipid in water emulsions. The result is low-fat margarines and other spread type products. In some instances, the polysaccharide seems to act as both emulsifier and stabilizer, but it should be emphasized that current interest is in systems where substances like pectin are used to stabilize emulsions which also requires the use of emulsifiers in the mixture. Pectin is not a very good emulsifier *per se.* A typical experimental approach is shown in Figure 2.9 where the pectin concentration required to stabilize an emulsion of soybean oil emulsified with sodium caprylate is illustrated (Kravtchenko et al., 1995). The stabilization concentration is defined as that required to reduce the flocculation rate by 90%. By using this sort of empirical approach, which is quite typical of this kind of investigation, results similar to those shown in Table 2.2 can be obtained. Since both emulsifiers and stabilizers probably exert their effects by surface interactions, complex effects are possible, and this is the explanation for the variable effectiveness of the different gums and stabilizers. Pectin is effective at much lower concentrations than dextrans, but it is not possible to say that this is simply because it is charged, since the explanations are likely to be more complicated. Different pectins can vary considerably too. Table 2.2 summarizes the various combinations that have been described. Although it is possible to characterize the emulsions in many ways – from semi-empirical settling studies to precise measurements of particle size

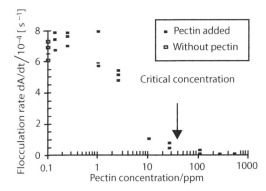

FIGURE 2.9   The left-hand axis shows an empirical measure of the flocculation rate of an emulsion of soybean oil and sodium caprylate, estimated from the rate of change of the absorbance. The lower axis shows the concentration of added pectin on a logarithmic scale. The filled squares show that pectin has a systematic effect on the settling rate. This is a typical empirical method by which the properties of different colloid stabilizers are compared.

TABLE 2.2   Hydrocolloid Effects on Emulsions

Degree of adsorption	Mechanism	Outcome
Strong adsorption of polymer	Steric hindrance to interactions	Stabilization
Incomplete, partial absorption	Bridging between particles	Flocculation, fast creaming
Non-adsorption	Promotes coalescence	Destabilize, fast creaming
In all cases flocculation, increased viscosity	No effect	Slower and creaming

Source: Adapted from Bergenstahl, B. et al., (1995), in Dickinson, E. and Lorient, D. (eds.), *Food Macromolecules and Colloids*, Royal Society of Chemistry, Cambridge, 1995.

distributions – it remains difficult to predict the effect of any particular commercial sample in other than the most general terms.

More complex systems still have been developed, where interactions between polysaccharides, like pectin, and proteins, such as whey proteins or gelatin, have been used to modulate the emulsifying effect. Pectins and gelatins can form stable gel-type structures, and in combination with their emulsifying and stabilizing ability on lipid emulsions are used to make low-fat spreads of growing commercial importance. These tend to need much higher concentrations than simple emulsions, and unantici-pated problems of macromolecular interactions have been encountered. A typical formulation is: water, sunflower oil, hydrogenated vegetable oil, caseinates, salt, modified starch, emulsifier, sodium alginate, vitamin E, potassium sorbate, vitamin A, flavourings, β-carotene and vitamin D, given as usual in descending order of amount but not specifying the exact concentrations. The emulsifier is usually a gum, though it may be a monoglyceride or phospholipid, while alginate is an alternative to pec-tate. Modified starch also has a stabilizer role. There is obviously plenty of scope for macromolecular interactions in such a mixture.

2.1.11   PHASE SEPARATION EFFECTS

As a result of the use of more complex mixtures including proteins and polysaccharides at higher concentrations some unexpected phase separa-tions have been encountered. These have, in fact, been well known for many years (Tompa, 1956) and were first accurately described for syn-thetic polymers in organic solvents. Since then, similar effects have been recognized in mixtures of biological macromolecules in aqueous solution (Albertsson, 1971). Most of the research has been done on mixtures of proteins and dextrans or hyaluronic acid (Edmond and Ogston, 1968) and, under conditions where the macromolecules are neutral or where charge effects are suppressed (by for instance using a high ionic strength sol-vent system), these are generally believed to involve coexclusion effects depending upon the volume occupied by the molecules. Above certain concentrations, known as the *critical point*, the mixture spontaneously separates into two phases, showing interestingly enough a water–water interface. Pectin and gelatin mixtures show these effects, and an example of a binodal plot is shown in Figure 2.10. Generally each phase is rich in one of the components, although both components are always present in both phases. It is possible to make dispersions and even emulsions of the one phase in the other, and this can be used to obtain spherical particles which can then be gelled. Also, added substances partition between the two phases, while colloidal sols tend to be concentrated into the interface. While, so far, no far-reaching applications for these two-phase systems have been developed it is possible that there will be a demand for highly characterized polysaccharides in the future as components of two- or even three-phase systems. The number of phases is not limited to two, and in principle it is possible to obtain an extra phase for each extra polymer. Systems with up to five phases have been described although in practice three is rarely exceeded.

For further discussion and examples of the use of two-phase systems of this kind refer to Harding et al. (1995). Since pectins are charged, and may be highly charged, and so is gelatin away from its isoelectric point (pH 4), incompatible coexclusion may give way to the rather different phenomenon of complex coacervation. In this, two phases appear but, unlike co-exclusion, both macro-molecules are present in the same phase, and hardly any in the other. The effect is due to charge–charge interactions and is seen where the gelatin has a net positive charge (below the pI) and the pectin has a negative charge. A high ionic strength can even so sometimes suppress these interactions sufficiently for incompatible exclusion to occur instead.

The problems for pectin production are more to do with specification of commercial samples. As we have seen, a crucial parameter for pectin is its DE and this is always specified. In addition, the molecular weight is probably significant although it is not always given. For more highly developed applications not only this but also the distribution of molecular weights or polydispersity and the use of the correct average for the weight is likely

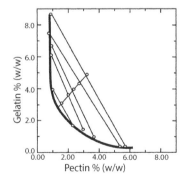

FIGURE 2.10   A binodal plot for gelatin–pectin mixtures in 0.5 M NaCl pH 5.5, 50°C. These conditions were chosen to minimize charge interactions and gelatin–gelatin interactions. The phase separation is the result of incompatible coexclusion. Two phases are present at all compositions within the curve, and only one outside it. The results are relevant to the formulation of low-fat spreads.

to become important. Indications are that the *number* average molecular weight (see Section 1.1.3) may be the one needed, and will have the best correlation with properties in the various applications for pectins.

In addition to the uses of pectins mentioned above, over the recent period the use of the pectins as immunomodulators and also other biological effects have been in focus. Structure activity relations are covered in a paper by Ferreira et al. (2015) and also in Chapter 3.

2.2 CELLULOSE AND HEMICELLULOSES

The basic structure of cellulose has already been described in Section 1.1.9. It is a (1→4) linked β-ᴅ-glucan:

$$...→4) \ β\text{-}ᴅ\text{-}Glc\textit{p}\text{-}(1→...$$

The *hemicelluloses* (Coughlan and Hazlewood, 1993) are a group of substances which occur associated with cellulose, and are usually a mixture of xylans or xyloglucans, arabinogalactans, glucomannans and galactoglucomannans.

Cellulose is the principal component of most plant cell walls, and is for the moment mainly of interest because of processes for paper making, and as a major structural component of textile fibres such as cotton (*Gossypium* spp.), ramie (*Boehmeria nivea*), jute (*Corcorus capsularis*), flax (*Linum usitatissiumum*) and sisal (*Agave sisalana*). It is worth mentioning here that 'surgical cotton' is actually made from wood. Kapok, a minor crop, derives from *Ceiba pentandra* and *Bombax malabaricum* and is of special interest because the lumen of the fibres is full of air, making it buoyant and useful in flotation devices.

There is little or no human food interest, since human digestive enzymes cannot break down the components of the cell wall. Ruminants, however, can use it with the aid of the rumen microflora and there is much current interest in processes such as silage making which involve bacterial fermentation of plant raw materials. The substrate is often grass and the cell wall materials are altered and made more accessible as animal feeds. There is also growing interest in the role of cellulose as an important part of the 'fibre' in the human diet which, while not absorbed, has second-order nutritional effects.

Physical properties of products where cellulose takes part have been studied recently and hydrogels containing cellulose have been thoroughly reviewed (Shen et al., 2016). The production of single cellulose chains and how to prepare these have been reviewed by Yuan and Cheng (2015), while Trovalli (2013) gives a review on the bacterial celluloses, an area which is developing fast.

The present uses, important as they are, might well be less significant than the potential, which is enormous. Cellulose is the most plentiful bioresource on earth, and is produced by the plants at the rate of around 4×10^{10} tonnes per year. The breakdown products are potential feedstocks for many chemical processes, but above all cellulose is a huge potential source of glucose, and thence of fuel ethanol, and there is extensive work under way designed to develop effective cellulases which may one day be used to break it down on a large scale. Waste paper may be the first economic substrate. Cellulases are also important enzymes involved in the growth of saprophytic organisms on cellulose substrates and justify consideration on that ground alone. On the other hand, the bacterium *Acetobacter xylinum* deposits pure cellulose fibrils on the surface of its culture medium. This, while obviously a very expensive source of cellulose, has been considered useful for special applications as an aid to skin grafting.

2.2.1 ENZYMATIC DEGRADATION OF CELLULOSE

Although cellulases have been available and studied for many years they have not as yet found widespread application. Two excellent reviews by Coughlan (1985) and Beguin (1990) give an account of the hundreds of studies aimed at obtaining these enzymes and understanding their mode of action. In the laboratory, they are used to assist in the disruption of plant cell walls, a typical use being the removal of the cell wall from yeast in some experimental methods. The cell wall contains many other components as well as cellulose, which prevents access to the substrate of even the most effective cellulase, while the insoluble and structured nature of cellulose fibrils also makes them a difficult substrate.

There is, however, a more fundamental problem in achieving complete hydrolysis of cellulose, which it has in common with all biological polymers. The fundamental structural unit of cellulose is the disaccharide cellobiose, although the fundamental residue is α-D-glucose. There are two ways in which hydrolytic attack might occur: by removal of the units at the end of the chain, or by fission of the chain in a random manner. Since biological polymers are asymmetric – that is, the chain has a direction and is different at each end – a single enzyme could only attack one end of each chain if it were of the exohydrolase type. Also it will remove the structural unit, which is often a disaccharide, while the ultimate requirement may be for the monosaccharide residue. This will require further hydrolysis by another enzyme acting on the disaccharide. This will inevitably be a very inefficient way of bringing about hydrolysis. For example, if the polymer contains 1000 units (and cellulose can have a DP of up to 14,000) and if the polymer is present at a concentration of $1 \, \text{g} \, \text{L}^{-1}$ the concentration of the ends will only be about $1 \, \text{mg} \, \text{L}^{-1}$ and in a structure such as the cell wall they may have very limited accessibility. Not only is the effective concentration very low, but it does not increase during the digestion of the chain. However, it could be argued that once the enzyme has made contact with the end of the chain it would probably remain in close contact, working its way along in an efficient manner. If it has to leave the chain and then make random contact again the rate of reaction must be very low. Thus, although cellulose is almost certainly synthesized in this way by stepwise addition to the end, though relatively slowly, this kind of hydrolysis is not feasible as a means of breaking down cellulose or any other biological polymer of any size.

Random attack on the bonds at any point in the chain will also be a poor method of obtaining complete breakdown. Apart from simple statistical considerations other factors are likely to influence the outcome. For example, the affinity of the enzyme for the substrate will probably decrease as the size of the fragment decreases, and it is common to find that hydrolysis becomes limited well before the fragments are down to single units.

Natural selection seems to have found the same answer to this problem for all the polymers, or at least all those that require breakdown as part of the digestive process, whether this is for bacteria, fungi or the human gut. Thus proteins, starches, nucleic acids and celluloses are all broken up by a combination of enzymes which act together both on the ends and randomly along the chains. If, as is the case for proteins, exo-enzymes capable of acting at either end of the chain are present, every time the main chain is split two sites of action for the exo-enzyme are formed and they become more effective as the fall in average chain length makes the endo-enzymes less effective. The same mechanism is found in starch degradation – as will be discussed in Chapter 4 – though in that case additional complications due to branching require extra enzymes for full breakdown.

It is difficult to think of a better way in which full breakdown of long-chain polymers could be achieved and it is interesting that suitable sets of enzymes have evolved, probably quite independently, for all the major biological polymers. A possible alternative mechanism could be imagined. If an enzyme containing several identical active sites had appeared, capable of aligning itself with a substrate polymer, this might have been sufficiently active to bring about efficient hydrolysis. No such enzyme has been described, though enzymes such as fatty acid synthetase with six different active centres on a single peptide chain are known. The most recent developments of biotechnology where two cellulases have been fused onto a single peptide chain may be the first step towards this kind of approach to the problem.

Thus enzymatic mechanisms have serious fundamental limitations when it comes to hydrolysis of polymers, and this is important since it both provides an explanation for the slow rate of progress towards a rather obvious and desirable target and must affect any estimate of the chances of future success. Cellulose is a very long polymer, with an aligned structure firmly held by hydrogen bonds, and its physical state significantly alters the rate at which it is attacked, as might be expected. This raises the question of preliminary physical treatments to open up the structures. The same considerations also come into starch processing, while native globular proteins are far more resistant to protease attack than unfolded ones. All these considerations make the study of cellulases rather different from the classical approach, which is effective for soluble enzymes acting on soluble substrates but which did not until recently develop methods for the study of combinations of enzymes acting on insoluble structured substrates.

We now consider the different classes of cellulases, and the way in which they interact.

2.2.2 ENDOGLUCANASES

These enzymes are able to hydrolyze amorphous cellulose, but have little activity on crystalline cellulose. They also hydrolyze oligosaccharides derived from cellulose and soluble derivatives like carboxymethylcellulose. The latter is used in a test of the randomness of the distribution of the fission sites. The ratio of the number of reducing ends released to the change in the viscosity of the solution can be used as a measure, since splits concentrated near the ends of the molecules will reduce the viscosity much less than those near the centre. The viscosity is sensitive to the distribution of chain lengths. Measurements of rates also showed that the longer fragments are preferred over the shorter ones as substrates. The ultimate product is cellobiose.

Tests of this kind show that many organisms secrete a mixture of endocellulases with slightly different specificities. For example, *Trichoderma viride* produces at least four versions of the enzyme with molecular weights around 50,000 and a carbohydrate content of about 10%. This suggests a glycoprotein of about 450 residues. Some caution is needed in interpreting earlier reports of diversity and carbohydrate content. Values of up to 50% carbohydrate have been reported but this probably represents contaminant carbohydrate rather than covalently bound oligosaccharide chains. Also, many reports of multiple enzymes are based on a fairly crude analysis of fungal or bacterial culture media, often by gel filtration methods. This can give highly misleading apparent molecular weights, as well as providing ample opportunity for protease attack on the enzymes. Two of the four enzymes preferred internal linkages while one of the others had an exo-action at the non-reducing end as well as attacking some internal links. Since the organism is actually producing both endo- and exo-enzymes this conclusion clearly depends on proof that there was no cross-contamination between the two classes of enzyme in the preparations. Again, in earlier work adequate proof of homogeneity was not always provided.

2.2.3 EXOCELLOBIOHYDROLASES AND β-GLUCOSIDASES

These enzymes cleave cellobiose from the non-reducing end of cellulose. They are just as widespread as the endohydrolases, and it is doubtful if one ever occurs without the other, except in deficient mutants. They are unable to attack carboxymethylcellulose, and are inhibited by cellobiose and glucose. There was some doubt whether they are made by bacteria, but it now seems likely that they are present as part of an enzyme complex. The glucosidases provide the final step in breakdown and are mostly concerned with the hydrolysis of cellobiose to glucose. They can, however, remove some glucose from the non-reducing end of cellulose fragments up to the hexose. *Trichoderma reesei* is a major source of cellulases for many applications but has relatively little β-glucosidase activity. This means that it can sometimes suffer from cellobiose inhibition, which can be relieved by the addition of extra β-glucosidase from another organism. This is a fairly obvious candidate for a gene transfer to enhance the ability of *Trichoderma* to produce β-glucosidase.

2.2.4 ENZYME COMPLEX AND CELLULOSOMES

Far more detailed information is now available in the form of some 50 sequences of various endo- and exo-cellulases and β-glucosidases (Ohmiya et al., 1997). In addition, the gene structure of many species has been analyzed. The glucanases fall into seven classes based on sequence analysis, with extensive homology evident between all the cellulases and xylanases from a wide range of microorganisms. The structure by x-ray crystallography of a cellobiohydrolase from *T. reesei* has been determined and serves as a prototype for the whole group. The enzymes have a catalytic domain, but often have another prominent and highly conserved sequence that corresponds to a cellulose binding domain. It has been known for many years that cellulases, unlike most proteins, do bind fairly strongly to cellulose. *T. reesei* has a highly conserved element of about 30 amino acid residues in all its cellulases, separated from the catalytic domain by a linker sequence rich in Pro-Thr-Ser residues and heavily glycosylated. When the binding domain was removed by papain treatment the catalytic domain retained activity against soluble substrates but lost most of it against cellulose. The binding region has been characterized by nuclear magnetic resonance (NMR) spectroscopy and appears to be well suited to binding between pairs of cellulose chains. Exocellulases have similar structures, and in this case the proteolytic fragment obtained from *Cellulomonas fimi* exocellulase was shown to bind to cellulose. A fusion of the binding region of this enzyme with the β-glucosidase of *Agrobacterium* has been made as well as one where an alkaline phosphatase was fused to a cellulose binding region. In the latter case, the idea was to use specific adsorption on a cellulose matrix as a means of purification of the alkaline phosphatase.

An important development of the last few years is the recognition that the enzymes not only have interlocking activities which are required to hydrolyze cellulose, but also that they are probably present in the form of complexes – thus ensuring that they are at the optimum position. Saprophytic bacteria and fungi can undoubtedly hydrolyze crystalline cellulose. Indeed, they can destroy plant cell walls. It is believed that they possess small organelles called cellulosomes (presumably similar to the well-known lysosomes found in most cells and known to contain packages of hydrolytic enzymes) which make contact with the surface of the plant, and contain the full package of cellulases. Since a layer of wax for example protects leaves, and bacteria are known to make a cutinase, which hydrolyzes it, there may be other hydrolases present as well. The complex present in *Clostridium thermocellum* contains about 18 different peptide chains, each of which may carry a distinct hydrolase activity. There are 15 endoglucanases, two xylanases, two β-glucosidases, one cellobiohydrolase and one lichenase. It also has a polypeptide called *CipA* which seems to act as a sort of scaffolding for the enzymes while having no enzyme activity itself. Non-covalent bonds of the kind that hold subunits together hold the complex together and has binding domains known as cohesins, which interact with domains called dockerins on the individual enzymes. These domains are highly conserved and are able to interact with calcium ions, which are involved in the complex formation. There may also be specific domains involved with anchorage of the complex to the cell surface beyond those already characterized as cellulose binding domains. This organism has at least 15 genes for endoglucanases, two for xylanases and two for β-glucosidases, suggesting that the multiple forms of the enzymes found in early work may not all have been due to post-translational modification or variable glycosylation.

Similar complexes have been found in other bacteria (*C. thermocellum* strain JM20 has up to 50 different peptide chains in the complex), but are not present in organisms such as *C. fimi* and *T. reesei* where the enzymes are secreted into the medium. It is known that the endoglucanases are synthesized with a leader sequence and tend to contain two or three introns, but the genes do not appear to be clustered and notably there is no evidence for polycistronic mRNA.

Whenever a complex consisting of several enzymes is found the question always arises as to whether the activities might be present on a single peptide chain. It is not always easy to answer this question. Thus, the fatty acid synthetase of chicken liver is made up of six activities all on a single chain of about 500 residues, but it is possible to break up the chain with proteases and retain all the activities on what are then separate peptides. In fact, unobserved fission during earlier work led at first to the belief that there were actually six different peptides until better methods led to their isolation in one piece. As against this the fatty acid synthetase of plants has, despite all efforts, not been isolated in one piece and does really seem to be seven individual peptides. That for yeast occurs on two peptide chains, which are actually coded by two genes on different chromosomes, so all possibilities can exist. The fact that only small pieces of mRNA were found, appropriate to the individual cellulose degrading enzymes, and that the genes appear to be fairly widely separated, suggests that they are not all on one chain. There is some evidence of clustering of endoglucanase genes, which may be the result of homologous recombination.

2.2.5 TRANSGENIC EXPRESSION

Most expression has been in *Escherichia coli* though many other hosts can also be used, notably *Saccharomyces cerevisiae* and *Bacillus subtilis*. As always the ability of the host to secrete the enzyme into the medium is of interest and this varied in an unpredictable way. Three different genes for endoglucanases from *B. subtilis* when expressed in *E. coli* were ~70, 24, and only 1% secreted although their sequences are almost identical, including the signal peptides. The latter do not seem to be involved in secretion in any case. The *E. coli*-produced enzymes are as usual not glycosylated, though again as usual those from yeast may be. *Bacillus* enzymes have found use in detergent formulations.

In a most interesting use of transgenic expression, Olsen et al. (1995) fused two β-glucanase activities onto a single chain. The β-glucans of barley are (1→3), (1→4) linked β-glucans with cellulose-like domains of about 15 consecutive glucose residues with (1→4) β links, separated by mainly (1→3) linked regions of about three residues. In all about 70% of the links are (1→4) β, and are attacked by typical (1→4) linked β-glucanases with the evolution of cellobiose. Barley, however, produces an enzyme which attacks the (1→4) β-link provided there is an adjacent β(1→3) link, and the action of this enzyme is important in the events of germination. The β(1→3), (1→4) glucanase is susceptible to thermal denaturation and is often insufficient in barley, leading to the appearance of β-glucans in the beer made from it, and also limits the nutritional value in animal feeds. Although thermostable hybrid versions of this enzyme are available and can be added they would not be very efficient at breaking down cellulose-like molecules to cellobiose or glucose since an endo-activity is really needed as well.

While investigating the carrot rot organism *Erwinia carotovora*, a popular organism for cellulase studies, the gene for an endocellulase was cloned. It proved to encode the catalytic domain only of the enzyme and lacked the cellulose binding domain and the linking peptide chain. It was nevertheless able to attack both a β-glucan and carboxymethylcellulose. The sequence was fused with that for a hybrid thermostable β(1→3), (1→4) glucanase from *Bacillus* and expressed in *E. coli*. This version failed to fold correctly, but another attempt in the yeast *Pichia pastora* was glycosylated and secreted and found to be active. Called *CELGLU*, the hybrid enzyme was able to degrade soluble cellulose and possessed both activities. That is, the tetrasaccharides produced by the (1→3), (1→4) glucanase were further broken down to glucose and a triose by the endocellulase activity. It was more effective than a simple mixture of the two enzymes.

2.2.6 FUSED ENZYMES

The reason why fused enzymes or multifunctional enzymes are more effective than an equivalent mixture of separate enzymes is not immediately obvious, but lies in the *kinetic basis of enzyme action.* Consider an enzyme with an ordinary small soluble substrate. The rate of reaction will depend on the collision frequency. If the active site occupies about 5% of the enzyme surface, and if the substrate is in the correct orientation when it makes contact about 10% of the time, then one collision in every 200 will result in fission. The average Brownian motion jump for a protein like an enzyme is less than one Ångstrom and the much higher mobility of the substrate is responsible for most of the collisions. This is why immobilized enzymes are in general just as active as soluble ones.

Now consider the situation with an insoluble large substrate like cellulose. The only motion is likely to be that of the enzyme, so that while the chance of a successful collision is about the same the frequency is much lower. There may be mechanisms which improve the chance of a successful orientation which involve sliding along chains once a successful contact has been made. The existence of the cellulose binding domain in cellulases of all kinds is almost certainly an adaptation to this situation, and its importance in improving the effective collision frequency is obvious. There is some evidence for a special form of binding in the action of amylase on starch (Chapter 4) and it might be expected in all enzymes involved in degrading polymers. Even so, the rate is likely to be inherently slower and anything that improves the chance of successful contact would be an advantage. This is probably what fused enzymes deliver.

The active centre for the next step is placed near the relevant substrate group, and must be better placed than if a separate enzyme has to make its way to the appropriate spot from anywhere in the system by a series of random collisions.

There seems little doubt that this kind of fused multienzyme is the way ahead for cellulose degradation. The role and need for several different but complementary activities is now understood, but experience suggests that simply adding these enzymes to the reaction mixture either together or sequentially is not likely to give useable rates of degradation, even of soluble or amorphous cellulose, still less of crystalline cellulose or intact cell walls. There can be no certainty that useful results will be obtained even with fused multienzymes, but they represent the best chance of reaching what remains a very attractive target. It should be noted that it is not necessary that all the enzymes are covalently linked. It is sufficient that they should remain in a close and fixed orientation during activity, which could be achieved by non-covalent subunit interactions.

2.2.7 SOURCES AND CURRENT USES OF CELLULASES

The modern era of cellulase study seems to date from the last war when the US army realized that its soldiers' uniforms were disintegrating rather rapidly in the jungles of the Pacific. A particularly virulent strain of *T. viride* was isolated from a cartridge belt (Coughlan, 1985) and a research programme started which continues to run. Another source of research activity is the minor industry devoted to the prevention of wood rot, which also involves cellulolytic activity. These are applications where cellulase activity is not wanted but there are others where cellulase activity was desired, such as the conversion of straw waste to cattle food.

An enormous range of organisms which seems to include most invertebrates are known to make cellulases, or at least use cellulases obtained from symbiotic microflora. One of the most remarkable is the teredo, a wood-boring bivalve otherwise known as the shipworm, which does great damage to wooden ships. It hosts large numbers of celluloytic bacteria which appear to be the source of the cellulase it uses in boring into the ships' timbers.

The principal organisms used for cellulase production are *T. reesei*, *T. viride* and *Erwinia* spp. Cellulase production is induced by cellulose and inhibited by cellobiose, and mutants of *Trichoderma* have been found which are very productive (30 g mixed cellulases per litre) and seem to make the full range of endo-and exo-cellulases. *Trichoderma* protoplasts can be transformed easily with plasmids and the very high production has been used in fused genes to make chymosin.

In processes for the saccharification of cellulose it has been calculated that the cost of the enzyme represents at least half of the overall costs, so an efficient producer will be crucial to the process. A major use for cellulases is the clarification and extraction of fruit juices and in some processes where partial removal of the cell wall is helpful in rehydration, as in dried soup manufacture. They have been investigated as a way of reducing case hardening in raisins: these become dehydrated when incorporated into some types of breakfast cereal, and fail to rehydrate fast enough when mixed with milk before consumption. This seems to be a simple barrier effect which could be eased by in effect making holes in the skins. Until recently it has not been possible to transform the organisms present in the rumen, so attempts to improve their celluloytic activity met with no success. A plasmid has now been isolated which apparently can do this, and in the near future attempts to shift rumen microflora will probably succeed. It would be especially useful if the abilities of the microflora against crystalline cellulose could be improved.

Cellulases find use in textile processing, as a means of loosening fibres, and have had some small applications as an aid to extraction of lipids from oilseeds. In most cases, a mixture of cellulases, xylanases and hemicellulases is used in very crude preparations. As yet there is no process for the production of glucose from cellulose or for the breakdown of straw and similar agricultural waste in full-scale operation. Starch is very well established for glucose production and major strides will be needed before cellulose is likely to replace it. Even so, a process for the complete degradation of cellulose remains one of the most attractive targets in the whole of biotechnology. It is not likely to be achieved in the near future and this may be an example where simply applying existing enzymes will not be sufficient. It will be necessary to devise new ones with more activity than any that occur naturally.

2.2.8 WASTE PAPER PROCESSING

The polysaccharide component of waste paper is generally a fairly pure cellulose, and it has long been seen as a good target for enzyme treatment. It is potentially a source of glucose but offers the possibility of recycling. Any use that could be found for it would also alleviate a major disposal problem. In the United States, about 15 million tonnes of old newsprint are potentially available for reprocessing, and the figure in Europe is similar. About half of this is actually recycled and one of the problems is the removal of ink. This involves pulping the paper with caustic soda, and a detergent and other reagents to release the ink, followed by washing, flotation and bleaching with hydrogen peroxide. The de-inking process itself causes considerable disposal problems. In some cases where water-based rather than oil-based inks are used the problems are different but also damaging to the environment.

Attempts have been made to use cellulases and hemicellulases to release the ink particles and they do work up to a point. Presumably, the enzymes attack the surface of the fibres and release ink particles by detaching the residues to which they are bound. Their use does reduce the hazards of the wash effluents.

Recently, a curious effect on the fragmentation of fibres has been discovered. When newsprint was treated with cellulases in a simple laboratory-scale reactor, the ink particles appeared to attach themselves to smaller fibres present in the newsprint with a relatively high lignin content. They could then be separated by a filtration process, giving two fractions, roughly equal in weight, one free of ink and one retaining all the ink. It is thought that when cellulase binds to cellulose the binding domain can bring about disruption of the cellulose structure without necessarily any accompanying hydrolysis, and it is possible that this effect of ink release is due to this. In this application that would be useful since it is desirable to maintain the fibres at a useful size, for re-use.

Cellulases have been inserted into tobacco, more specifically cellulases and xylanases from *C. thermocellum*, and have been observed to have some effect on the status of the cell wall in tissue culture when expressed at high level. The idea is twofold. Forage containing high levels of endogenous cellulase might be easier to digest when eaten by domestic animals, and it might also be easier to make paper from the appropriate source plants. Cellulases have also been expressed in barley, and are expected to improve malting characteristics.

2.3 XYLANS AND XYLANASES

Xylan is the third major component of cell walls, found at 15%–30% in hardwoods, 7%–10% in softwoods and up to 30% in the stems of plants such as maize and ramie. It is also found in some species of seaweed. It is mostly amorphous, shorter than cellulose and has branches containing glucuronyl (α-D-Glc$p$A) and arabinosyl (α-L-Ara$f$) residues. The glucuronyl residues confer on it polyanionic properties. Xylan may be covalently linked to other cell wall components, especially lignin, and forms an interface between cellulose and lignin. Its backbone is based on the pentose D-xylose unit linked by β (1→4) bonds:

$$\ldots \to 4) \; \beta\text{-D-Xyl}p\text{-}(1 \to \ldots$$

It is only partially soluble in water, but its soluble part has been characterized (see Dhami et al., 1995 and references cited therein) and Table 2.3 summarizes some fundamental properties. Determination of its conformation has been difficult: aggregation phenomena have obscured interpretation of, for example, the Mark–Houwink viscosity a coefficient, but it does on the basis of non-ideality phenomena appear to adopt a highly expanded asymmetric coil type conformation in solution.

Over the years, xylans have been shown to have a great variety of uses. Ebringerova (2012) has written an extensive overview of the potential that the xylans have as biomaterial resources, and her conclusion is that xylans have both food and non-food applications. The use in food ranges from functional food ingredient over to importance in bread making and as

TABLE 2.3   Some Fundamental Properties of a Water Soluble Xylan

Property	
Polymer type	Polyanionic; branched
Conformation type	Zone C (semi-flexible/asymmetric coil)
Solubility	Only partially soluble in aqueous solvents
Molecular weight	~150 kDa (water-soluble fraction)
Intrinsic viscosity [η]	~42 mL g^{-1} (water-soluble fraction)

additives in ice-cream making, and as a dietary fibre. The potential in production of nanoparticles for use as a drug carrier in pharmaceuticals is also interesting. Additionally, various biological functions have been recorded during the last decade, where both immune-enhancing properties, management of wounds and antioxidant effects have been shown. Da Silva et al. (2012) describe xylans as a promising hemicellulose for pharmaceutical use in the future. New excipients of biological origin will be very interesting in the future. As the xylans has undergone investigations as functional food it was discovered that the oligosaccharides produced by xylans had a great potential as a prebiotic material and also proved to be effective (Singh et al., 2015).

Xylanases on the other hand are very widely found in microorganisms of all kinds (Prade, 1996). Not surprisingly degradation requires the action of several enzymes working in conjunction and xylanases are part of the cellulosome complex discussed above, when it is present.

Most organisms produce two or three endoxylanases which cleave the xylan to oligosaccharides and eventually xylobiose, the disaccharide which is then cleaved to xylose by a β-xylosidase. Unlike cellulases there do not appear to be any exoxylanases. Other enzymes may also be present that attack the side chains, such as arabinosidases and glucuronidases. Interestingly, deletion of neither the gene for an endopolygalacturonidase and a xylanase in *Cochliosolus carbonum,* nor the gene for a cutinase in *Nectria haematococca,* had any effect on pathogenicity. It is possible that some of the other enzymes in the cellulosome compensated for the absence of these activities, since it is difficult to believe that the cellulosome is not involved in penetration of the host cell wall. Xylanases have many similarities with cellulases – for example, the same lysozyme-like mechanism – but do not show significant sequence homology. They also have multiple genes for the enzyme. In one interesting case, *Fibrobacter succinogenes,* the gene codes for a peptide bearing two xylanase catalytic domains with two different specificities. This was shown by splitting the two domains by proteolysis, and then examining the pattern of xylobiose, tetraose and pentaose produced, and other multiple xylanases may also have subtle specificity differences. Other examples are *Neocallimastix patriciarum* where a gene coding for an endoglucanase, a cellobiohydrolase and a xylanase has been found. *Ruminococcus flavefaciens* actually secretes a single peptide with both xylanase and lichenase activity. In *Caldocellum saccharolyticum* no less than five open reading frames coding for a xylanase, a β-glycosidase, an acetylesterase and two other unidentified enzymes have been found. An arabinofuranosidase activity has also been found in similar circumstances, and all the enzymes involved in degrading both the main chain and the side chains of xylan have been described.

It is tempting to speculate that fused genes of this type are on the way to the kind of multiple activity discussed above with all its potential advantages, but it remains likely that different organisms do have systems perhaps representing different evolutionary stages starting with distinct enzymes through complexes of subunits to multiple activities on a single chain. It is worth noting that in all these cases the minimum molecular weight capable of supporting an activity seems to be about 25,000 corresponding to about 200 amino acid residues, although of course many enzymes, including lysozyme itself, are only about half this size.

The nature of the enzyme mix, as well as the presence or absence of a complex, is clearly something to take into account when choosing the source organism for enzymes for applications. It may one day be thought unfortunate that so much use has been made of *Trichoderma* since it appears not to have a complex enzyme at all.

Like cellulases, xylanases have a non-catalytic binding domain, called the xylan-binding domain. The crystallographic structure of *T. reesei* xylanase has been determined and shows a cleft that might be a binding site, as does a xylanase from *Streptomyces lividans*. The latter, however, shows differences and there may be two distinct classes of xylanase structure. The structure of a typical bacterial xylanase is shown in Figure 2.11 and also shows the way in which the xylan lies in a cleft.

In vitro mutagenesis has been used to introduce extra cysteine groups and intrachain disulphide bonds into *Bacillus circulans* xylanases. The objective was an increased thermal stability, for which there is a clear requirement in paper processing usage. It was only partly successful, since it was not possible to improve on the stabilities already available from naturally occurring enzymes. Finally *E. coli*, with additional genes coding for pyruvate decarboxylase and ethanol dehydrogenase from *Zymomonas mobilis*, is able to use xylose for ethanol production. Other organisms have also been transformed and it is now possible to envisage a cell wall breakdown process yielding both glucose and xylose which could then both be fermented to make fuel ethanol. Breakdown of lignocellulosic material yields hexoses and pentoses in a 6:4 ratio, and there would not be much point in only being able to use the hexose. In degradation of wood the initial process is usually steam heating or grinding which results in an intitial breakup of the xylan to shorter chains. Xylanases can then be used to hydrolyze the chains further, but so far this is not sufficiently efficient for large-scale commercial use.

Recent reviews of the xylanase production basically in enzymes also give details related to the specificity and uses of the different xylanases that have been isolated (Motta et al., 2013; Juturu and Wu, 2014).

FIGURE 2.11   The structure of α-xylanase from *Bacillus subtilis*. In contrast to the pectin lyase illustrated in Figure 2.4 the xylanase is built almost entirely from helices, with a β-sheet structure surrounding the active centre. The binding of the substrate chain is also shown. (Reprinted from Pickersgill, R. et al., in Geisow, M.J. and Epton, R. (eds), *Perspectives in Protein Engineering*, Mayflower Worldwide Ltd., Birmingham, 1995.)

2.3.1 TEXTILE USES

Cotton is possibly the most important crop with an annual value of about $20 billion, and has naturally attracted some attention. Classical plant breeding has led to considerable improvement in the crop over the years, but it has lost a large part of its market to synthetic fibres. The opportunities offered by transgenic cotton are seen by the industry as a way of recovering some of this lost market share.

The cotton fibre itself is under the control of a large number of genes and it has not been possible to do much to improve the fibre by conventional breeding. There are two diploid Asiatic varieties (*Gossypium arboretum* and *Gossypium herbarium*) and two main tetraploid American varieties (*Gossypium hirsute* and *Gossypium barbadense*). A long-term breeding target has been to find a variety with low gossypol levels, but when one was found it could not be used because the fibre was not acceptable. Gossypol is a toxic polyphenolic substance found in cotton seed which severely restricts the usefulness of the seed left after cotton fibre removal, and this is one issue which could well be revisited with modern methods. If gossypol could be removed from commercially successful varieties it would transform its value as a cattle food, though experience with other species suggests that unexpected effects may also occur. Removal of similar compounds from lupins revealed that they were insect repellents, and the crop was devastated, and something similar might be anticipated with gossypol.

Agrobacterium tumefaciens has been used as a vector for cotton transformation for several years, but this is a comparatively limited method since only a few varieties of cotton can be regenerated, the number of genes that can be inserted by this method is limited, and they take about 14 months to form a transgenic plant. Particle bombardment methods are now in use, and can produce transformed plants more quickly. While genes believed to control fibre quality are being investigated, the results are so far subject to commercial secrecy. However, they do include genes controlling giberrellins, and also the gene complex for the production of polyhydroxybutyrate and other alkanoates. The target is to induce the formation of the polyhydroxyalkanoate as part of the cotton fibre, perhaps in the lumen, which should result in very different fibre properties. There are two different types of fibre on the cotton seed: the short lint, which can be absent in some varieties, and the longer more valuable fibres with lengths between 15 and 40 mm and a diameter of 12–25 μm. At maturity the lumen is empty, and the fibres are nearly all cellulose, with only 0.9% pectin and waxes, and 1% or 2% protein. It should be fairly easy to define the required improvements in these fibres, and at least some of the characteristics, like length, are under the control of single genes.

Although almost the whole of the economic value of cotton is in the fibres the potential of the remaining seed should not be overlooked in the list of targets for genetic engineering methods. Some of the cotton cellulose is used in the chemical industry.

Xylanases are used to facilitate the isolation of textile fibres from soft plants such as flax and ramie. These stems have a lower lignin content than woods, and the lignin can be removed more easily than in paper making (see Section 2.3.2). Even so, it can cause difficulties and in some processes is liable to browning reactions. Enzymatic treatment is under less severe conditions and leads to good quality fibrils with little staining. This might be the base of some traditional methods which may well involve bacterial xylanases. Even so, the method is relatively undeveloped, and the mechanisms are not fully understood (Prade, 1996).

2.3.2 PAPER MAKING

Paper is made from wood pulp, and is largely cellulose fibres, though it may also carry fillers such as china clay. Most of it is made from conifers, though poplars and eucalyptus are important temperate and tropical sources with eucalyptus now accounting for 15% of world pulp production. The main problem in paper making is to separate the cellulose fibres which make up about 50% of the mass, from lignin and other components such as pectins, xylans and elastins. This is achieved by heating in alkaline conditions, and the addition of sulphite or hypochlorite as oxidizing agent. Covalent links between cellulose and lignin require powerful oxidizing conditions for their rupture, while residual lignin gives a brown colour to the paper and requires a bleach. The lignin and waste chemicals are usually disposed of as effluent, a practice which has been increasingly questioned, and will inevitably become more expensive in the future.

Abnormal lignins are known in a number of adventitious mutants in maize (*Zea mais: Sorghum bicolor*), rice (*Oryza sativa*) and yams. In addition rubbery wood, a disease of apple trees, is known to involve abnormal lignin formation. Recently, attempts have been made to alter the lignin in trees suitable for paper making by suppressing the production of cinnamyl dehydrogenase which is involved in the final stages of lignin synthesis. The gene has been identified in eucalyptus and poplar. The effect of suppression is to make it easier to extract the lignin by reducing the covalent linking in the cell wall. This has two beneficial effects. Less disruption is needed in making the pulp and releasing the lignin, which increases the yield and reduces degradation of the cellulose. In addition, less residual lignin means that less bleach is required, and the effluents are easier to handle. Genetically modified poplars and eucalyptus are now being grown in sufficient quantity for trials of pulping, and it is likely that as new trees can be phased into the paper making cycle – which could take up to 30 years for the slower growing species – paper making will be significantly affected by this development. Since eucalyptus requires only 7 years to reach useful maturity there may actually be a shift in the type of trees used.

In a separate development, attempts are being made to use enzymes to assist in the separation of the components of the cell wall. In the course of Kraft pulping, in which the pulp is heated in alkaline conditions, the hemicelluloses and the lignin are dissolved and partly degraded, but eventually the pH drops and the xylan and lignin reprecipitate onto the cellulose fibrils. The lignin goes brown as a result of oxidation of this polyphenol, and it has been found that bleaching can be improved by the addition of xylanases. These break up the xylan, releasing reducing sugars, and making it easier to remove the lignin, which may appear in the form of soluble lignin–carbohydrate complexes. It is obvious that cellulase activity in the xylanases would be undesirable, since good paper requires the preservation of the fibril structure as much as possible.

Thus the source of the xylanase is of some interest. It is possible to fractionate the enzymes in the typical fungal preparation, though this is in practice much too expensive for a commercial enzyme, and it is possible to inhibit the cellulase selectively by using proteases. However, the best way appears to be the use of cellulase-free mutants, and possibly by genetically engineered strains. The relatively high temperatures and pH of the medium in which the enzyme is required to act provide plenty of scope for improvements on the existing enzymes.

2.4 THE 'GUM' MANNANS: LOCUST BEAN GUM, GUAR AND KONJAC MANNAN

The mannans (commonly referred to as 'gums' along with other industrial polysaccharides such as xanthan) are found in numerous seeds located mostly in the cell wall and can be regarded as cell wall components, though they are mobilized on germination in much the same way as storage polysaccharides. As a group they have traditional uses as adhesives, hence the name, and more recently as food components.

The mannans represent an important group of neutral and water-soluble (albeit with varying degrees of difficulty) polysaccharides based on the repeating mannose group linked β (1→4). Their importance – particularly locust bean gum – has been for use in foods although, as with many of the other polysaccharides, they are finding interest in the pharmaceutical and biomedical areas. Pure polymannan:

$$...→4) \text{ β-D-Man}p(1→...$$

is structurally very similar to cellulose, and like cellulose, presumably through hydrogen bond associations of the chains, is insoluble. Although present in ivory nuts, it is very rare and of no commercial value. However, two naturally occurring variants of this – with galactose side chains – are of particular importance: these are the galactomannans known as *locust bean gum* and *guar*, extracted from the seeds of leguminous plants and which have monosaccharide galactose side chains to variable degrees (locust bean gum < guar):

$$...→4) \text{ β-D-Man}p(1→4) \text{ β-D-Man}p(1→...$$
$$6$$
$$\uparrow$$
$$1$$
$$\text{α-D-Gal}p$$

Locust bean gum is sometimes called carob gum, because it comes from the endosperms of the leguminous Mediterranean tree carob (*Ceratonia siliqua*), whereas guar comes from the seeds of *Cyamopsis tetragonoloba* found mainly on the Indian subcontinent. One of the principal effects of the galactosyl substitution is to increase solubility, presumably by disrupting interchain hydrogen bonding: guar, with the higher degree of substitution, is the more soluble.

Like alginates – as we see in Chapter 5 – galactomannans are represented by a 'G:M' ratio (M:G ratio preferred for alginates) although the G and M have different meanings: in locust bean gum G stands for galactose and M for mannose (whereas in alginate G is guluronic acid and M is mannuronic acid). The G:M ratio for guar is 0.5–0.6 (i.e. ~1:2) whereas for locust bean gum it is much lower (G:M = 0.2–0.3 or ~1:4). Another galactomannan, *tara gum*, has an intermediary value (G:M = 0.33 or ~1:3).

The distribution of the G side chains along the M–M backbone is not random: with locust bean gum the substituted residues are clustered together into groups of about 25, leaving long stretches of virtually free polymannan (and providing ideal interaction sites with the seaweed polysaccharides carrageenan and agarose (Chapter 5) and xanthan (Chapter 6)). In guar, the distribution is also irregular but without the long polymannan stretches found in locust bean gum, making it less potent for interactions with other polysaccharides.

Another important variant on the polymannan theme is the glucomannan *konjac mannan* extracted from the tubers of *Amorphophallus konjac*: with glucomannans there are no side chains but the glucose residues are found interspersed (to varying degrees) inside the polymannan chain:

$$...\rightarrow 4)\ \beta\text{-}D\text{-Man}p(1\rightarrow 4)\ \beta\text{-}D\text{-Glc}p(1\rightarrow 4)\ \beta\text{-}D\text{-Man}p(1\rightarrow ...$$

In konjac mannan approximately one residue in every 5–6 is acetylated, which is enough to confer water solubility.

One of the principal biotechnological interests in galactomannan and glucomannan molecules at the moment is their behaviour in mixed polysaccharide systems (Figure 2.12). There is particular interest in the synergistic interactions between both galactomannans and glucomannans with xanthan, carrageenans and agarose: mixtures do give enhanced gelation and other properties but the precise mechanism of interaction (viz. do the mannans interact with the so-called 'ordered' or 'disordered' forms of xanthan) is currently under considerable debate. Because subtle differences in the way these polysaccharides interact has proved to be significant in food uses, a new process for the enzymatic modification of guar gum to make it behave in some respects like locust bean gum has been introduced, as we consider in some detail in Section 2.4.1.

Today, the mannans have obtained a variety of uses and due to that enzymatic degradation is also important for the production of the mannans with specific properties. A minireview by Moreira and Filho (2008) describes the degrading enzyme systems of importance.

2.4.1 LOCUST BEAN (CAROB) GUM

Locust bean gum comes from the endosperm which represents ~40% of the seed of the carob tree, and is associated with John the Baptist ('St John's

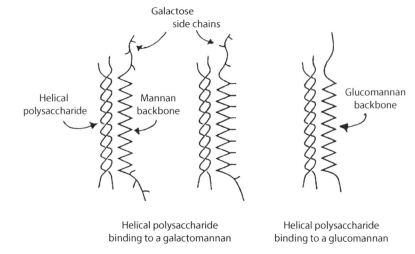

Galactose
side chains

Helical
polysaccharide

Mannan
backbone

Glucomannan
backbone

Helical polysaccharide
binding to a galactomannan

Helical polysaccharide
binding to a glucomannan

FIGURE 2.12   Early model proposed for the interaction between helix forming polysaccharides and either galactomannans or glucomannans. This early model is now taken to be too simplistic and is the subject of considerable research using structural (x-rays), rheological and hydrodynamic methodologies. (From Morris, V.J, In Harding, S.E., Hill, S.E. and Mitchell, J.R. [eds], *Biopolymer Mixtures*, Chapter 14, Nottingham University Press, Nottingham, 1995, and reproduced with permission). For stimulating discussions on this topic the reader is referred to this article and that of Morris, E.R. (1995).

Bread' and 'locusts and wild-honey'). More importantly from a commercial standpoint is its usefulness primarily in foods and health food products. The carob tree grows in Spain, Italy, Cyprus and other Mediterranean lands, although more and more tree-growing areas are being lost to urban development and it is becoming a more expensive and less attractive proposition.

2.4.1.1 EXTRACTION

The fruit pods from the carob seeds are harvested, cleaned and broken. The recovered seeds are rolled, split and the endosperms separated by differential grinding. The endosperm is then milled, blended and graded with respect to the viscosity of the gum. Preparations with impurities from the rest of the seed tend to degrade faster because of the higher enzyme content.

2.4.1.2 PROPERTIES

Table 2.4 summarizes some of the fundamental properties of locust bean gum. Both the molecular weight (as determined from sedimentation equilibrium in the ultracentrifuge) and the intrinsic viscosity are typical for other polysaccharides such as alginate and carrageenan, although the solution conformation – because of the lack of charge repulsion effects – is much more flexible. Solution viscosities are generally invariant over a

TABLE 2.4   Fundamental Properties of Locust Bean Gum

Property	
Polymer type	Neutral
Conformation type	Zone C–D (semi-flexible/random coil)
Mark–Houwink a coefficient	~0.8
G:M ratio	~0.2–0.3
Solubility	Depends on G:M ratio and temperature
Molecular weight	~350 kDa
Intrinsic viscosity [η]	~1000 mL g^{-1}
Gelation properties	Needs cross-linker such as borax, or the synergistic presence of κ or ι- carrageenans, agarose or xanthan (very strong gels)
Key sites for biomodification	Polymannan regions; binding by helical regions of other polysaccharides

wide range of pH (3–11). Besides degradation from seed enzyme impurities, bacterially contaminated samples are also – like other plant polysaccharide gums – liable to chain degradation.

Locust bean gum is only partially soluble in water and can be fractionated on the basis of this water solubility: Hui and Neukom (1964) found that ~30% was soluble in cold water with progressively increasing amounts dissolving with increase in temperature, and reported that the cold water-soluble (CWS-lbg) material had a higher G/M ratio than the material soluble only in hotter conditions (HWS-lbg). The additional suggestion that the hot water-soluble fraction also had a considerably higher molecular weight was not however supported by later measurements (McCleary et al., 1985; Gaisford et al., 1986) which showed that both the molecular weights and intrinsic viscosities were approximately similar for the two fractions.

2.4.1.3 GELATION

Locust bean gum will gel in the presence of cross-linkers like borax. Even more potent, and attractive from a food-grade acceptability, is the synergistic gel formation in the presence of a second polysaccharide. The GM-free stretches of polymannan are *thought* to bind the helical regions of carrageenans, agarose and xanthan, and are terminated when a galactosyl-M residue is encountered, encouraging the next polymannan sequence on the locust bean gum chain to interact with a different molecule in the mixture: in this way a network can be readily formed. We return to this in Chapters 5 and 6.

Changing the ratio of locust bean gum to the second polysaccharide gives great control on the strength and syneresis properties of the gel. These gels are usually prepared by heating to 85°C (to ensure solubilization of the locust bean gum) before allowing to set. Another popular modification to the locust bean gum structure is etherification to give hydroxyethyl-locust bean gum or carboxymethyl-locust bean gum: these are particularly useful in the textile industry for sizing and printing.

2.4.1.4 FOOD USES

Locust bean gum has particularly favourable hydration/swelling or water binding properties compared with other polysaccharides, making it ideal as a stabilizer in ice cream, cheese spread and other dairy products with a high water content hydration/swelling ability (locust bean gum content can range from 0.05% to 0.6%, and is often used in conjunction with other (cheaper) polysaccharides such as carboxymethylcellulose and κ-carrageenan).

Health and dietary products: Along with guar, pectins and other gums it also promotes D-glucose tolerance and appears to be effective at lowering blood lipid content.

Pharmaceutics and cosmetics: Interest is primarily in the interaction with xanthan, and patents have been issued with regards to their usefulness in sustained-release hydrophilic matrix systems and as a direct compression excipient for hydrophilic matrix tablets (Melia, 1991). Like xanthan and carrageenan it is also used as a toothpaste thickener.

2.4.2 GUAR

The decline in supply of locust bean gum from its Mediterranean habitat has to some extent been offset by the increased production of guar from the leguminous plant *Cyamopsis tetragonolobus* from the Indian subcontinent (which produces ~90% of commercial guar) where the monsoon rain pattern provides ideal guar producing conditions.

Guar extraction: As with locust bean gum, guar is contained in the endosperm part of the seed and a similar extraction procedure is followed. The seeds are separated from dirt, stones etc. and then split, with the protein-rich germ part of the seed removed by a cyclone separator or something equivalent, and then a sifter. After a soaking step (optional) and temperature treatment (not above ~110°C where the guar becomes heat unstable) the endosperms are ground/dehusked, washed, hydrated, removed of particulates and then air dried to give a fine milled but molecularly intact guar.

2.4.2.1 PROPERTIES

Table 2.5 summarizes some of the fundamental properties of guar. Guar is a relatively high molecular weight (~1×10^6 Da), neutral polysaccharide and like locust bean gum has a fairly flexible conformation in solution. Because of the larger number of G side chains it is more soluble than locust bean gum: in the laboratory, dissolution (this also applies to locust bean gum) is normally facilitated by wetting with a small amount of ethanol (0.5 mL ethanol per 100 mg guar) to prevent any risk of aggregation, dispersion in distilled water for about 5 minutes using a mixer and then heating for 10 minutes at ~80°C followed by cooling. Viscosities are very high, but guar solutions, like locust bean gum, exhibit a strong shear (thinning) dependence and strong (reversible) temperature dependence.

TABLE 2.5   Fundamental Properties of Guar

Property

Polymer type	Neutral
Conformation type	Zone C–D (semi-flexible/random coil)
G:M ratio	~0.6
Solubility	Higher than locust bean gum
Molecular weight	800–1000 kDa
Intrinsic viscosity [η]	~1200 mL g^{-1}
Gelation properties	Needs cross-linker such as borax, or the synergistic presence of κ or ι- carrageenans, agarose or xanthan (weaker gels than locust bean gum systems)
Key sites for biomodification	Polymannan regions: binding by helical regions of other polysaccharides; OH groups for esterification and etherification

2.4.2.2 GELATION

Like locust bean gum, guar will form relatively weak or fluid gels in the presence of cross-linking agents such as transition metal ions and borates, the latter requiring high alkalinity since the pK_a of boric acid is as high as 9.3 and they are self-healing.

2.4.2.3 CHEMICAL MODIFICATION

As with locust bean gum – where the problem is greater – the relatively poor solubility of native guar has inspired the production of more soluble derivatives. This has been achieved by etherification (Prabhanjan et al., 1989) of the hydroxyl groups to give the polyanionic derivative carboxymethylguar (substituent group: $-CH_2COO^-$, usually the Na$^+$ salt, degree of substitution, DS ~ 0.05–0.3) and the neutral derivative hydroxypropylguar ($-CH_2CH(OH)CH_3$, DS ~ 0.3–1.6). These increase the solubility of the guar without affecting guar's naturally viscous properties. A polycationic derivative is 2-hydroxy-3-trimethylammonium chloridepropylguar ($-CH_2CH(OH)CH_2-N^+$ $(CH_3)_3Cl^-$, DS ~ 0.05–0.2).

2.4.2.4 ENZYMATIC MODIFICATION OF GUAR GUM

Guar is generally readily available from an annual harvest. This is in contrast to the situation for many other gums which are now available on a less reliable basis and are often produced in small and fluctuating amounts. In a very ingenious application of enzyme technology it has been found possible to convert guar into a gum closely resembling locust bean gum by treating it with an enzyme. Locust bean gum is produced mainly around the Mediterranean and for political reasons and as agriculture changes its supply has become unreliable. On the other hand, it is widely used in food applications because it has particularly desirable properties. Both gums are galactomannans, but guar contains around 38%–40% galactose as side chains while locust bean has only 22%–24% galactose. This is enough to alter their properties significantly. The mannan backbone is quite rigid, giving high viscosity in solution, while galactomannans with fairly low galactose content such as locust bean gum give useful elastic gels when mixed with other polysaccharides such as xanthans and carrageenan. The high galactose gums do not do this so effectively, and are preferred when the aim is simply to raise the viscosity. Initial trials with a galactosidase obtained from Lucerne showed that an enzyme could be used to modify the galactose content.

Guar itself has an α-galactosidase which appears on hydration and is probably involved in the mobilization of the carbohydrate reserves. It can remove galactose side chains from the galactomannan and produce something with rheological and gelling properties near to those of the locust bean gum at a similar galactose content. The process could not be used, however, for two reasons. First, the amount of enzyme present in the aleurone cells of the bean was about 1% of the protein at the most, and it could not be considered as a viable commercial source of the galactosidase. Second, other enzyme activities were present. A β-mannoside mannohydrolase and a β-mannanase, which attacked the main mannan chain, were also active and made the product useless. It would be hard to imagine a better example of the advantages of transgenic production of an enzyme since both of the main advantages apply, that is, a secure source of the enzyme and an absence of contamination. The seed mRNA route was followed (Overbeeke et al., 1989), and when cloned led to 7 out of 3000 cultures expressing the enzyme. This is somewhat less than would have been expected from the known abundance of the enzyme. When guar seeds germinate there is a surge in the mRNA for the hydrolytic enzymes generally and this can lead to the appearance of high levels of the enzymes. The mRNA levels subsequently fall and there is a temporal disjunction between the two levels which is probably responsible for this effect. Similar effects are also known for amylases. Sequencing of the cDNA showed that the mature enzyme has 364 residues and a molecular weight of 39,777 Da, in reasonable agreement with the value of 40,500 Da obtained from SDS gels. In fact the difference is enough to accommodate six amino acid residues, but taking into account the likely error of the SDS determination the difference is not cause for concern. The enzyme has an extra 47 residues based on the mRNA, and certainly has a pre-sequence, probably involved in targeting to the endosperm polysaccharide. The sequence is longer than that previously reported for amylases of barley and wheat, and might be a pre-pro-sequence. However, unlike amylase the induction of the enzyme is not giberrellic acid dependent, and may have a quite different secretion mechanism. It is probably glycosylated. The sequence of the guar, human and yeast enzymes are shown in Figure 2.13 and have a considerable degree of homology. The enzyme endogenous to *E. coli* is different, with poor homology and also differs in requiring Mn^{2+} and NAD^+, and it is possible that there are at least two different classes of galactosidases.

The guar enzyme has now been successfully expressed in yeast, which provides a source of it. It should be added perhaps that it is more effective on guar gum than the constitutive yeast enzyme itself. The enzyme, like so many of those that function on polysaccharides, is able to debranch the gum at about a 30% solids level, a concentration at which the mass

```
1     6    11   16   21   26   31
LONGL ARTPT MGWLH WERFM CNLDC QEEPD SCISE KLFME MAELM VSEGW KDAGY EYLCI DDCWM          Human
     +***  **  ***   *  *  *          **  _+ * *+ ** *+   ** *+ , ****
1                                    23
AENGL GQTPP MGWNS WNHFG CD           INE NVVRE TADAM VSTGL AALGY QYINL DDCWA           Plant
     ***  *  *   ***    ** *  *           * +++ + *** +   **  +** ** * ****
1                                    23
SYNGL GLTDQ MGWDN WNTFA CD           VSE QLLLD TADRI SDLGL KDMGY KYIIL DDCWS           Yeast

66                                  101
APQRD SEGRL QADPQ RFPHG IRQLA NYVHS KGLKL GIYAD VGNKT CAGF- PGSFG YYDID AQTFA DWGVD     Human
     +** ***  +    ** * *+ **  **** ***** *_* * ** *    ***+_* + * * *** ****
56                                  91
ELNRD SEGNM VPNAA AFPSG IKALA DYVHS KGLKL GVVSD AGNQT CSKRM PGSLG HEEQD AKTFA SWGVD     Plant
     ** *+ + *    ** * + +** +*    - + *** ** * *    **** ** * ** **
56                                  90
SG-RD SDGFL VADEQ KFPNG MGHVA DHLHN NSFLF GMYSS AGEYT CAGY- PGSLY REEED AQFFA NMRVD     Yeast

135                          166
LLKFD GCYCD SLENL ----- ADGYK ----H MSLAL NRTGR SIVYS -CEWP LYMWP FQKPN YTEIR QYCNH WRNFA   Human
     +**+* * ***    +*     * ** +** *++* *** + +      + * **
126                          156
YLKYD NC--- --ENL ----- GISVK ERYPP MGKAL LSSGR PIFFS MCEWG WEDPQ IWAKS ----- -IGNS WRTTG   Plant
     ***** **        +*     * ** +** ***+* +* ** +  +*         * * ** +*
124                          159
YLKYD NCYNK GQFGT -PEIS YHRYK ----A MSDAL NKTGR PIFYS LCNWG QDLTF YWGSG ----- -IANS WRMSG   Yeast

200                              225     235
DIDDS WKSIK ----- ----- }----- ---SILDW TSFNQ ERIVD VAGPG GWNDP DMLVI GNFGL SWNQQ VTQMA LWAIN AAPLF MSNDL   Human
     **+* * *+           ** *        +**** ***** *** * ** *+ +   + + +**+ ***+ + *+
185                              206     216
DIEDN WNSMT ----- ----- ----- ---SIADS NDKWA S---- YAGPG GWNDP DMLEV GNGGM TTEEY RSHFS IWALA KAPLL VGCDI   Plant
     * +            **   * *       ** * **** * *** * + + +* + *** ***+ * **+* +
188                              231     240
DTAEF TRPOS RCPCD GDEYD CKYAG FHCSIMNI LNKAA PMGQN -AGVG GWNDL DNLEV GVGNL TDDEE KAHFS MWAMV KSPLI IGANV   Yeast

270                          305
RHISP QAKAL LQDKD VIAIN QDPLG KQGYQ LRQGD ----- NFEVW ERPLS GLAWA VAMIN RQEIG GPRSY TIAVA   Human
     * *     * + ***+* ** ** ** ++           +*** *** +* * ++*
251                          286
RAMDD TTHEL ISNAE VIAVN QDKLG VQGKK VKSTN ----- DLEVW AGPLS ANKVA VILWN RSSSR ATVTA SWSDI   Plant
     +  ++ + +*+* ***+* ** +  +         ++ +* ***   * *+* *      +  + +*
275                          320
NNLKA SSYSI YSQAS VIAIN QDSNG IPATR VWRYY VSDTD EYGQG EIQMW SGPLD NGDQV VALLN GGSVS RPMNT TLEEI   Yeast

340                  375
SLGKG VACNP ACFIT QLLPV KRKLG FYEWT SRLRS HINPT GTVLL QLENT MQMSL KDLL     Human
     * *           * *++ + * * **
321          353
GLQQG TTVDA RDL-W EHST- -QSLV SGEIS AEIDS HALKM -YVLT PRS     Plant
     * +     *        * +
355                  390                            425
FFDSN LGSKK LTSTW DIYQL WANRV DNSTA SAILG RNKTA TGILY NATEQ SYKDG LSKND TRLFG QKIGS LSPNA ILNTT VPAHG IAFYR LRPSS   Yeast
```

FIGURE 2.13 Homologies between yeast, human and the guar α-galactosidase. The sequences have been aligned and identical residues are marked with an asterisk, while similar residues are marked +. Guar shows a 53% homology with the human enzyme, and slightly less with yeast. (Reprinted from Overbeeke, N. et al., *Plant Mol. Biol.*, 13, 541–550, Figure 4. With permission.)

is an opaque firm gel. In developing the process, trials with a variety of levels of gum were used. In general high water contents are undesirable because of subsequent drying costs, so the aim was to obtain the highest solid content consistent with effective enzyme action. The results shown in Figure 2.14 indicate that at 50°C an effective removal of galactose takes place at about 30% guar, and that this is actually an optimum concentration. The way in which enzymes interact with long-chain polymers is still little understood, but evidently the rules that would be expected to apply to simple soluble substrates are not a good guide for the behaviour of polysaccharides.

As pointed out in the general discussion of enzyme use, purified enzymes are rarely if ever used and it has been found in practice that

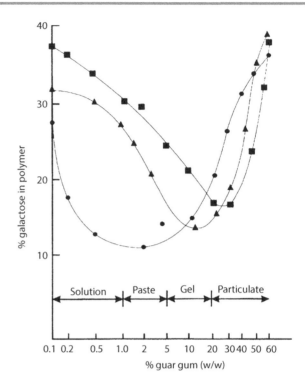

FIGURE 2.14 The galactose content remaining in the polymer plotted against the concentration of guar gum, after reaction with α-galactosidase for 24 hours. Reaction was at 35°C (●) 42°C (▲) and 50°C (■). The physical state of the mixture is also indicated. At 50°C there is optimum activity at about 30% guar in a particulate gel. (Reprinted from *Carbohydrate Polymers*, 12, Bulpin, P.V. et al., Development of a biotechnological process for the modification of galactomannan polymers with plant α-galactosidase, 155–168, 1990, with permission from Elsevier.)

a very convenient method is to use the disrupted producing organism as a sort of matrix. In many instances, what appears to happen is that the enzyme becomes attached to the cell wall, usually cellulose based, and functions as in effect an immobilized enzyme. In a development, Schreuder et al. (1996) describe ways in which the cell wall mannoproteins can be used to ensure that enzymes made by heterologous expression in yeast can be targeted to the wall. The structure of the yeast cell wall is indicated in Figure 2.15 showing the position of the α-agglutenin, which is the mannoprotein. In a most ingenious application, a fused gene was made containing the C terminal half of the mannoprotein, and the sequence coding for amongst other enzymes the α-galactosidase of interest from guar. This was expressed in the yeast and covalently incorporated into the cell wall, where it retained its galactosidase activity. In a control experiment, the galactosidase expressed by the same yeast but without the fusion to the agglutenin was secreted but did not adhere to the cell wall. A cutinase and a lipase have also been successfully treated in the same way and there seems to be no reason why this approach should not be extended to a variety of enzymes of interest in

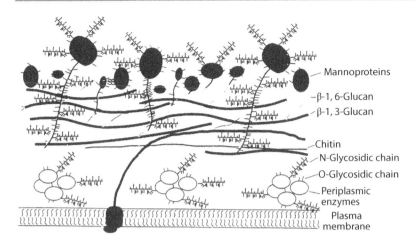

FIGURE 2.15   The structure of the cell wall of yeast *Saccharomyces cerevisiae*. The cell wall lies outside the plasma membrane, and consists of layers. The innermost is made up of chitin and glucans, and is cross-linked. Between it and the membrane is a collection of the 'periplasmic enzymes'. The mannoproteins that form the anchorage points for enzymes on the surface are covalently linked to the glucan layers. In the application of interest a fused gene producing both the mannoprotein and a galactosidase led to the appearance of enzyme mannoprotein fused products attached to the outer layer. (Reprinted from *Tibtech*, 14, Schreuder, M.P. et al., Immobilising proteins on the surface of yeast cells, 115–120, 1996, with permission from Elsevier.)

the polysaccharide field. Glucose isomerase is used in the form of a disrupted producing organism, while other possibilities in the use of yeast bearing suitable enzymes on the surface for clean-up effluent purposes can be anticipated.

2.4.2.5 REDUCED AND LOW-FAT PRODUCTS

In the last few years, there has been growing interest in low calorie versions of traditional spreads such as butter and margarine. The mouthfeel of mixtures of microcellulose and guar gum has been very important to this development. Guar mixed with cellulose and blended under suitable moisture conditions forms a surface layer, apparently based on hydrogen bonding, which alters the colloidal stability of the microcellulose suspension and leads to flocculation. Drying then produces the ingredient which can be blended with lipid and water to form spreads with reduced lipid content. There seems to be a specific interaction between guar and cellulose which is also used in paper treatments as a finishing step.

2.4.2.6 USES OF GUAR

These take advantage of the high viscosity, thickening and high shear thinning (without shear degradation of the polymer chain) properties (important in suspension technology) of guar and its high viscosity derivatives. The presence of small amounts of guar or its derivatives (0.001%–0.05%) can suppress turbulent flow in piping operations by greatly altering the drag (i.e. friction) coefficient. These properties are proving particularly valuable for well-bore technology in the oil industry and also in other mining operations as flocculants and flow aids.

2.4.2.7 FOOD USES

Like locust bean gum, guar is popular in dairy products because of its thickening and stabilizing properties together with its spreadability and is particularly popular in ice cream and cheese spread products. It is often used in conjunction with other polysaccharides like locust bean gum and carrageenan and has particularly favourable hydration/swelling or water binding properties compared with other polysaccharides, making it ideal as a stabilizer in ice cream, cheese spread and other dairy products with a high water content hydration/swelling ability (locust bean gum content can range from 0.05% to 0.6%). It is often used in conjunction with other (cheaper) polysaccharides such as carboxymethylcellulose and κ-carrageenan.

2.4.2.8 TEXTILE AND PAPER TECHNOLOGIES

Guar is becoming more popular for dye thickeners for application to carpets and fabrics, not only because of its thickening ability but also because of the ease of its removal associated with its shear thinning properties. In the paper industry, guar and particularly its cationic derivatives are used to increase the dry strength, drainage and retention properties of the cellulosic paper fibres.

2.4.2.9 PHARMACEUTICS AND COSMETICS

As with locust bean gum, the interest is primarily in the interaction with xanthan and patents have been issued with regards to their usefulness in sustained-release hydrophilic matrix systems and so on. Cationic guar derivatives are becoming popular as hair shampoo and hair conditioner thickeners.

2.4.3 KONJAC MANNAN

Konjac mannan is a glucomannan with a G:M ratio (G for glucose) of ~0.66 or 2:3. It is acetylated at positions C2 or C3 on the mannose residues to a total polysaccharide degree of substitution (DS) ~0.1. Although it is generally accepted that the G residues are intrachain with the M residues (rather than branched like galactomannans) there are conflicting views as to whether there is a regular repeat within the chain or whether it is a more or less random copolymer structure.

2.4.3.1 EXTRACTION

Konjac mannan is extracted from the corms of *A. konjac.* (The corms themselves – called 'devil's tongue' or 'elephant's foot' because of their shape – are used in several traditional Japanese foods.) It is produced commercially by harvesting and pulverizing the corms into a powder (konjac flour) which is air classified to separate konjac mannan from starch and other products.

2.4.3.2 PROPERTIES

Table 2.6 summarizes the fundamental properties of konjac mannan. Molecular weight from sedimentation equilibrium analyses is similar to that of locust bean gum although the intrinsic viscosity is somewhat higher. What is clear though is that konjac mannan solutions, like those of the galactomannans, are very prone to degradation by enzyme impurities: the presence of proteases or prior heat treatment (most enzymes denature irreversibly on heating) is important in characterization experiments.

Konjac mannans do not gel unless they are de-esterified by alkali treatment to remove the acetyl groups, to give gels which are thermoirreversible. Much stronger gels are produced by the synergistic interactions with helical polysaccharides, and as discussed above, this is of considerable interest at the present time.

Konjac mannans have also been used against obesity, and an overview of six clinical trials showed that the konjac mannans have potential as a bodyweight reduction remedy; studies did not prove a reduction in BMI (body mass index), but the number of trials and their length in time were not sufficient for drawing conclusions on this (Zalewski et al., 2015).

It is also relevant to mention in this overview that various glucomannans also have shown bioactivities of various kinds, and those from konjac as well as from another plant, *Bletilla striata*, have properties making them interesting as materials for development of drug delivery products as well as wound healing agents. The potential for the drug delivery properties is based on the strong gelling abilities as well as the fact that they are biocompatible (Wang et al., 2015).

TABLE 2.6   Fundamental Properties of Konjac Mannan

Property

Polymer type	Neutral: interrupted repeat
Conformation type	Zone C–D (semi-flexible/random coil)
G:M ratio	~0.66
Molecular weight	300 kDa
Intrinsic viscosity [η]	~1000 mL g^{-1}
Gelation properties	Needs alkaline conditions to remove acetyl groups. Gel thermoirreversible
Key sites for biomodification	Polymannan regions: binding by helical regions of other polysaccharides; OH groups for esterification and etherification

2.5 GUM EXUDATES

Although traditionally used in foods, these are particularly useful as adhesives in pharmaceutical creams: karaya gum has been used for many years for dentures. These adhesive properties derive directly from their biological roles in trees and shrubs for sealing and protecting damaged bark and so on. Their commercial use is however declining, largely because of the cost of harvesting and the advance of alternatives such as starch, cellulose derivatives and xanthan, and also because of sometimes undesirable yellow colouration deriving from absorption of tannins from bark. We nevertheless give the three principal ones – arabic, tragacanth and karaya – brief consideration here. Exudate gums often have a small (<4%) covalently attached protein.

2.5.1 GUM ARABIC

Gum arabic is a branched polysaccharide which derives from, and is hand-picked from, the exudate of *Acacia* trees with Sudan being the principal producer. Its backbone consists largely of $\beta(1{\rightarrow}3)$-linked galactose residues:

$$...{\rightarrow}3) \text{ } \beta\text{-}{\rm D}\text{-}{\rm Gal}p(1{\rightarrow})...$$

with intrachain rhamnose, arabinose and, especially near the ends, glucuronic or methylglucuronic residues. The side chains, which start with a $\beta(1{\rightarrow}6)$ Gal$p$ residue, are 3–5 residues in length and can also have these other residues present. The result of all this branching means that gum arabic can adopt a compact spheroidal conformation in solution ('zone E'), though if the number of glucuronate or methylglucuronate residues is larger then the conformation adopted will be more of an expanded coil ('zone D'). Total molecular weight is of the order of 350,000–500,000 Da. Its natural adhesive function in the plant has been taken advantage of commercially in its use (at ~40% concentration) for paper and card. In foods, it is used primarily to prevent sucrose crystallization in sweets and toffees. In pharmaceuticals, gum arabic suspensions help to stabilize salts of heavy metals – as used in calamine lotion – and also stabilize cod-liver oil.

2.5.2 GUM TRAGACANTH

This derives from hand harvested exudate from *Astragalus* spp. plants, chiefly in Iran, Syria and Turkey. Like arabic it is highly branched, but consists of two forms, one neutral (~40% by weight), the other (~60%) acidic. The neutral form ('tragacanthin') has a backbone consisting largely of $\beta(1{\rightarrow}6)$-linked galactose residues:

$$...{\rightarrow}6) \text{ } \beta\text{-}{\rm D}\text{-}{\rm Gal}p(1{\rightarrow}...$$

to which 1→2, 1→3 and 1→5 linked L-arabinose side-chain residues are attached. The other, acidic fraction ('tragacanthic acid') consists of a backbone principally of galacturonic acid:

$$...{\rightarrow}4)\ \alpha\text{-}{\rm D}\text{-}{\rm Gal}p{\rm A}(1{\rightarrow}...$$

also with side chains containing galactose, xylose, fucose and rhamnose. The total molecular weight of the structure has been determined by sedimentation methods to be of the order of 840,000 Da (Gralen and Karrholm, 1950).

Gum tragacanth has been of greater pharmaceutical interest than gum arabic, because of its greater heat stability and acid resistance and good emulsifying properties. It has found use for assisting the suspension of powders, as a base for jelly lubricants, in spermicidal gellies and in penicillin preparations. In foods, advantage has been taken of its long shelf life, and it has received the most use in salad dressings and sauces.

2.5.3 GUM KARAYA

This comes from the exudates of the north Indian bushy tree *Sterculia urens*, and is again hand harvested. Like tragacanth its molecular weight has again been determined by ultracentrifuge methods from the Uppsala laboratory, and found to be very large, almost 10 million daltons (Kubal and Gralen, 1948). Like arabic and tragacanth it is also a complex branched polysaccharide consisting of galactose, rhamnose and also galacturonate and guluronate residues, the latter two being in the Ca^{2+} and Mg^{2+} salt forms. In pharmaceutical applications, besides being used as a laxative, the classical use of gum karaya has been as a denture adhesive and is normally applied by dusting onto the binding surface of the denture prior to insertion. It is also used in powdered form in some food applications like sherbets and French dressings.

Gum karaya has, due to its range of functionalities, for a long time been of industrial interest, and recently a new area has been discovered, and this is as a drug delivery material. It is biocompatible and has been shown to be a good material for dosage forms for oral controlled realease of drugs (Setia et al., 2010).

2.6 GLYCOSAMINOGLYCANS

These are of major structural importance in vertebrates and besides hyaluronic acid ($M \sim 0.5$–2×10^6 Da) which is not sulphated, include chondroitin sulphate ($M \sim 20,000$ Da) and keratan sulphate ($M \sim 12,000$ Da) from connective tissue and dermatan sulphate from skin. A principal function is to form a structural matrix for the protein components of connective tissue and skin. In cartilage, for example, these molecules can form chondroitin sulphate proteoglycans: these are giant aggregates ($M \sim 10^7$–10^8 Da) formed from a hyaluronic acid backbone with

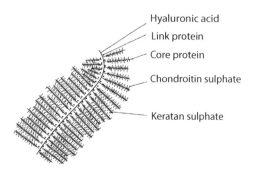

FIGURE 2.16   The complex structure of a proteoglycan. The hyaluronic acid polysaccharide backbone is non-covalently linked to core proteins which themselves have covalently attached chains of chondroitin sulphate or keratan sulphate. In cartilage this whole network provides a framework for binding and keeping collagen fibres in a strong and tight network. (Reprinted from *Dynamics of Connective Tissue Macromolecules*, Rosenberg, L., 107, 1975, with permission from Elsevier.)

covalently attached protein and then further chains of chondroitin or keratan sulphate covalently bound to the protein (Figure 2.16) and form a network for collagen fibres to bind and function. The size distribution of these huge aggregates has been investigated by light scattering techniques (Jamieson et al., 1992). Another pair of sulphated glycosaminoglycans are heparin and heparan sulphate, important natural coagulation factors (e.g. in blood clotting).

All these glycosaminoglycans have a linear 'alternating repeat' type of structure (Figure 1.5), one of the repeat units being either *N*-acetylglucosamine or *N*-acetylgalactosamine. All these substances used to be referred to as 'mucosubstances' or 'mucopolysaccharides' although this nomenclature has now been largely dropped to avoid confusion with mucus glycoproteins which line respiratory, alimentary and reproductive systems: these latter are *not* glycosaminoglycans and are based on a quite different structural principle.

2.6.1 HYALURONIC ACID

This is probably the most biotechnologically interesting of the glycosaminoglycans. Besides being found in the joints, umbilical cord and eyes (vitreous humour) of vertebrates it is also produced as a capsular component by the bacteria *Staphylococcus* and some *Streptococci*. Like the other glycosaminoglycans its primary structure is that of a linear alternating repeat. In this case the repeating residues are *N*-acetylglucosamine with glucuronic acid:

$$...\rightarrow3)\ \beta\text{-D-Glc}p\text{NAc}\ (1\rightarrow4)\ \beta\text{-D-Glc}p\text{A}\ (1\rightarrow...$$

X-ray diffraction studies reveal a helical structure, the precise nature of which (pitch etc.) is seemingly dependent on the type of counterion used. As mentioned, molecular weight is of the order of 1 million and preparations – from either animal or bacterial sources – are usually very polydisperse (see Figure 1.8a). The Mark–Houwink a (~0.8, Cleland and Wang, 1970; Gamini et al., 1992) and c coefficients (~0.6, Figure 1.8b) appear to suggest a somewhat stiffish coil (zone C) conformation in solution.

Its natural properties are becoming an increasingly important tool in biomedicine for the treatment of diseases involving joints such as arthritis and synovial fluid deficiencies. Its natural presence in the vitreous humour of eyes is taken advantage of in eye surgery (Stamper et al., 1990) and, like alginates and chitosans (Chapter 5), has been considered in wound healing technologies.

Hyaluronan has been shown to exhibit a great variety of biological functions as described by Dicker et al. (2014) and has received increased attention. The biocompatibility has led to the development of various drug delivery systems as well, and has proven to be valuable as a delivery system for anticancer drugs as well as for nucleic acid delivery, and can be used with great advantage in wound healing and against ailments like arthritis and ortheoporosis. Details related to this can be found in reviews by Dosio et al. (2016) and Khan et al. (2013).

FURTHER READING

Burton, R.A., Gidley, M.J. and Fincher, G.B. (2010) Heterogeneity in the chemistry, structure and function of plant cell walls, *Nature Chemical Biology*, 6, 724–732.

Lapasin, R. and Pricl, S. (1995) *Rheology of Industrial Polysaccharides. Theory and Applications*, Chapters 1 and 2, London: Blackie.

Pettolini, F.A., Walsh, C., Fincher, G.B. and Bacic, A. (2012) Determining the polysaccharide composition of plant cell walls, *Nature Protocols*, 7, 1590–1607.

Sousa, A.G., Ahl, L.I., Pedersen, H.L., Fangel, J.U., Sørensen, S.O. and Willats, W.G.T. (2015) A multivariate approach for high throughput pectin profiling by combining glycan microarrays with monoclonal antibodies, *Carbohydrate Research*, 409, 41–47.

Taylor, R.J., Welsh, E.J. and Morris, E.R. (1983) *Carbohydrates*, London: Unilever.

Pectin

Ferreira, S.S., Passos, C.S., Madureira, P., Vilanova, M. and Coimbra, M.A. (2015) Structure-function relationships of immunostimulatory polysaccharides: A review, *Carbohydrate Polymers*, 132, 378–396.

Rolin, R. (1993) Pectin, in Whistler, R.L. and BeMiller, J.N. (eds) *Industrial Gums*, Chapter 10, New York, NY: Academic Press.

Walter, R.H., (ed) (1991) *The Chemistry and Technology of Pectin*, New York, NY: Academic Press.

Celluloses

Batt, C.A. (1991) Biomass, in Moses, V. and Cape, R. (eds) *Biotechnology, the Science and the Business*, Chapter 28, London: Harwood Academic Publishers.

Shen, X., Shamshina, J.L., Berton, P., Gurau, G. and Rogers, R.D. (2016) Hydrogels based on cellulose and chitin: Fabrication, properties and applications, *Green Chemistry*, 18, 53–75.

Trovalli, E. (2013) Bacterial cellulose, in Dufresne, A., Thomas, S. and Pothan L.A. (eds) *Biopolymer Nanocomposites: Processing, properties, and Application*, pp. 339–365, New York, NY: John Wiley & Sons.

Yuan, X. and Cheng, G. (2015) From cellulose fibrils to single chains: Understanding cellulose dissolution in ionic liquids, *Physical Chemistry*, 17, 31592–31607.

Hemicelluloses

Coughlan, M.P. and Hazlewood, G.P. (eds) (1993) *Hemicellulose and Hemicellulases*, London: Portland Press.

Xylans and xylanases

Da Silva, A.E., Marcelino, H.R., Gomes, M.C.S., Oliveira, E.E., Nagashima, T., Jr, and Egito, E.S.T. (2012) Xylan, a promising hemicellulose for pharmaceutical use, in Verbeek, J. (ed) *Products and Application of Biopolymers*, Croatia: InTech Europe.

Ebringerova, A. (2012) The potential of xylans as biomaterial resources, in Habibi, Y. and Lucia, L.A. (eds) *Polysaccharide Building Blocks; A Sustainable Approach to the Development of Renewable Biomaterials*, New York, NY: John Wiley & Sons.

Juturu, V. and Wu, J.C. (2014) Microbial exo-xylanases: A minireview, *Applied Biochemistry and Biotechnology*, 174, 81–92.

Motta, F.L., Andrade, C.C.P. and Santana, M.H.A. (2013) *A Review of Xylanase Production by Fermentation of Xylan: Classification, Characterization and Application*, Croatia: InTech Europe.

Prade, R.A. (1996) Xylanase from biology to biotechnology, in Tombs, M.P. (ed) *Biotechnology and Genetic Engineering Reviews*, Vol. 13, pp. 101–32, Andover, UK: Intercept.

Singh, D.R., Banerjee, J. and Arora, A. (2015) Prebiotic potential of oligosaccharides: A focus on xylan derived oligosaccharides, *Bioactive Carbohydrates and Dietary Fibers*, 5, 19–30.

Mannanases

Moreira, L.R.S. and Filho, E.X.F. (2008) An overview of mannan structure and mannan-degrading enzyme systems. *Applied Microbiology and Biotechnology*, 79, 165–178.

Locust bean gum and Guar

Maier, H., Anderson, M., Karl, C. and Magnuson, K. (1993) Guar, locust bean, tara, and fenugreek gums, in Whistler, R.L. and BeMiller, J.N. (eds) *Industrial Gums*, Chapter 8, New York, NY: Academic Press.

Konjac mannan

Wang, Y., Liu, J., Li, Q., Wang, Y. and Wang, C. (2015) Two natural glucomannan polymers, from Konjac and *Bletilla*, as bioactive materials for pharmaceutical application, *Biotechnology Letters*, 37, 1–8.

Winwood, R. (1987) The konjac mannan/carrageenan mixed gelling system, PhD dissertation, University of Nottingham, UK.

Zalewski, B.M., Chmielewska, A. and Szajewska, H. (2015) The effect of glucomannan on body weight in overweight or obese children and adults: A systematic review of randomized controlled trials, *Nutrition*, 31, 437–442.

Synergistic interactions

Morris, E.R. (1995) Polysaccharide synergism—More questions than answers, in Harding, S. E., Hill, S.E. and Mitchell, J. R. (eds) *Biopolymer Mixtures*, Chapter 13, Nottingham: Nottingham University Press.

Morris, V.J. (1995) Synergistic interactions with galactomannans and glucomannans, in Harding, S.E., Hill, S.E. and Mitchell, J.R. (eds) *Biopolymer Mixtures*, Chapter 14, Nottingham: University Press.

Exudate gums

Setia, A., Goyal, S. and Goyal, N. (2010) Applications of gum karaya in drug delivery systems: A review on recent research, *Der Pharmacia Lettre*, 2, 39–48.

Whistler, R.L. (1993) Exudate gums, in Whistler, R.L. and BeMiller, J.N. (eds) *Industrial Gums*, Chapter 12, New York, NY: Academic Press. (See also his chapter in the 1973 edition of this book for a more extensive review.)

Hyaluronic acid

Ciba Foundation Symposium 143 (1989) *The Biology of Hyaluronan*, New York, NY: John Wiley & Sons, and articles therein.

Dicker, K.T., Gurski, L.A., Pradhan-Bhatt, S., Witt, R.L., Farach-Carson, M.C. and Jia. X. (2014) Hyaluronan: A simple polysaccharide with diverse biological functions, *Acta Biomaterialia*, 10, 1558–1570.

Dosio, F., Arpicco, S., Stella, B. and Fattal, E. (2016) Hyaluronic acid for anticancer drug and nucleic acid delivery, *Advanced Drug Delivery Reviews*, 97, 204–236.

Khan, R., Mahendhiran, B. and Aroulmoj, V. (2013) Chemistry of hyaluronic acid and its significance in drug delivery strategies: A review, *International Journal of Pharmaceutical Sciences and Research*, 4, 3699–3710.

SPECIFIC REFERENCES

Albertsson, P. (1971) *Partition of Cell Particles and Macromolecules*, 3rd Edition, New York, NY: John Wiley & Sons.

Bacic, A., Harris, P.J. and Stone, B.A. (1988) Structure and function of plant cell walls, in Stumpf, P.K. and Cohn, E.E. (eds) *The Biochemistry of Plants. A Comprehensive Treatise*, Vol. 14, Chap. 8, New York, NY: Academic Press.

Beguin, P. (1990) Molecular biology of cellulose degradation, *Annual Reviews Microbiology*, 44, 219–248.

Bergenstahl, B., Faldt, P. and Malmsten, M. (1995) in Dickinson, E. and Lorient, D. (eds) *Food Macromolecules and Colloids*, Cambridge: Royal Society of Chemistry.

Bulpin, P.V., Gidley, M.J., Jeffcoat, R. and Underwood, D.R. (1990) Development of a biotechnological process for the modification of galactomannan polymers with plant α-galactosidase, *Carbohydrate Polymers*, 12, 155–168.

Cheetham, P.J. (1995) Biotransformations—New routes to food ingredients, *Chemistry and Industry*, (April edition), 265–268.

Cleland, R.L. and Wang, J.L. (1970) Ionic polysaccharides III. Dilute solution properties of hyaluronic acid fractions, *Biopolymers*, 9, 799–810.

Coughlan, M.P. (1985) Fungal and bacterial cellulases and their production and application, in Tombs, M.P. (ed) *Biotechnology and Genetic Engineering Reviews*, Vol. 3, pp. 39–109, Andover, UK: Intercept.

Dhami, R., Harding, S.E., Elizabeth, N.J. and Ebringerova, A. (1995) Hydrodynamic characterisation of the molar mass and gross conformation of corn cob heteroxylan AGX, *Carbohydrate Polymers*, 28, 113–119.

Edmond, E. and Ogston, A.G. (1968) An approach to the study of phase separation in ternary systems, *Biochemical Journal*, 109, 569–576.

Gaisford, S.E., Harding, S.E., Mitchell, J.R. and Bradley, T.D. (1986) A comparison between the hot and cold soluble fractions of two locust bean gum samples, *Carbohydrate Polymers*, 6, 423–442.

Gamini, A., Paoletti, S. and Zanetti, F. (1992) Chain rigidity of polyuronates: Static light scattering of aqueous solutions of hyaluronate and alginate, in Harding, S.E., Sattelle, D.B. and Bloomfield, V.A. (eds) *Laser Light Scattering in Biochemistry*, Chapter 20, Cambridge: Royal Society of Chemistry.

Gralen, N. and Karrholm, M. (1950) The physicochemical properties of gum tragacanth, *Journal of Colloid Science*, 5, 21–36.

Harding, S.E., Hill, S.E. and Mitchell, J.R. (eds) (1995) *Biopolymer Mixtures*, Nottingham, UK: Nottingham University Press.

Hui, P.A. and Neukom, H. (1964) Properties of galactomannans, *Tappi*, 47, 39–42.

Jamieson, A.M., Blackwell, J., Zangrando, D. and Demers, A. (1992) Quasielastic and total intensity light scattering studies of mucin glycoproteins and cartilage proteoglycans, in Harding, S.E., Sattelle, D.B. and Bloomfield, V.A. (eds) *Laser Light Scattering in Biochemistry*, Chapter 18, Cambridge: Royal Society of Chemistry.

Kravtchenko, T.P., Parker, A. and Trespoey, A. (1995) Colloidal stability and sedimentation of pectin stabilised acid milk drinks, in Dickinson, E. and Lorient, D. (eds) *Food Macromolecules and Colloids*, Cambridge: Royal Society of Chemistry.

Kubal, J.V. and Gralen, N. (1948) Physicochemical properties of karaya gum and locust-bean mucilage, *Journal of Colloid Science*, 3, 457–471.

Lea, A.G.H. (1995) Enzymes in the production of beverages and fruit juices, in Tucker, G.A. and Woods, L.F.J. (eds) *Enzymes in Food Processing*, 2nd Edition, Chapter 7, Glasgow: Blackie.

McCleary, B.V., Clark, A.H., Dea, I.C.M. and Rees, D.A. (1985) The fine structure of carob and guar galactomannans, *Carbohydrate Research*, 139, 237–260.

May, C.D. (1990) Commercial sources and production of pectins, in Phillips, G. and Williams, P.A. (eds) *Gums and Stabilisers for the Food Industry*, Vol. 5, pp. 223–232, Oxford: Oxford University Press.

Melia, C.D. (1991) Hydrophilic matrix, sustained release systems based on polysaccharide carriers: Critical review. *Therapeutics: Drug Carrier Systems*, 8, 395–421.

Morris, V.J. (1987) New and modified polysaccharides, in King, R.D. and Cheetham, P.S. (eds) *Food Biotechnology*, Vol. 1, Chapter 5, London: Elsevier.

Ohmiya, K., Sakka, K., Karita, S. and Kimura, T. (1997) Structure of cellulases and their application, in Tombs, M.P. (ed) *Biotechnology and Genetic Engineering Reviews*, Vol. 14, Chapter 13, Andover, UK: Intercept.

Olsen, O.K., Thomsen, K., Weber, J., Duus, J.O., Svendsen, I., Wegener, C. and Von Wettstein, D. (1995) Transplanting two unique β-glucanase catalytic activities into one multienzyme which forms glucose, *Biotechnology*, 14, 71–74.

Overbeeke, N., Fellinger, A.J., Toonen, M.Y., Van Wassenaar, D. and Verrips, C.T. (1989) Cloning and nucleotide sequence of the α-galactosidase from *Cyamopsis tetragonoloba* (guar), *Plant Molecular Biology*, 13, 541–550.

Pickersgill, R., Harris, G. and Jenkins, J. (1995) Architecture and function of *Bacillus subtilis* pectate lysase and *Pseudomonas* xylanase A, in Geisow, M.J. and Epton, R. (eds) *Perspectives on Protein Engineering,* Birmingham, UK: Mayflower Worldwide.

Prabhanjan, H., Gharia, M.M. and Srivastava, H.C. (1989) Guar gum derivatives, part I. Preparation and properties, *Carbohydrate Polymers,* 11, 279–292.

Rosenberg, L. (1975) Structure of proteoglycans, In Burleigh, M. and Poole, R. (eds) *Dynamics of Connective Tissue Macromolecules*, Amsterdam, the Netherlands: North Holland.

Schreuder, M.P., Mooren, A.T., Toschka, H.Y., Verrips, C.T. and Klis, F.M. (1996) Immobilising proteins on the surface of yeast cells, *Tibtech,* 14, 115–120.

Science Daily (25 July 2016) Salad days: Tomatoes that last longer and still taste good. http://www.sciencedaily.com/releases/2016/07/160725122139.htm

Stamper, R.L., DiLoreto, D. and Schacknow, P. (1990) Effect of intraocular aspiration of sodium hyaluronate on postoperative intraocular pressure, *Opthalmic Surgery,* 21, 486–491.

Tompa, H. (1956) *Polymer Solutions,* London: Butterworth.

Uluisik, S., Chapman, N., Smith, R., Poole, M., Adams, G.G., Gillis, R.B., Besong, T.M.D., Sheldon, J., Stiegelmeyer, S., Perez, L., Fisk, I.D., Yang, N., Baxter, C., Rickett, D., Fray, R., Blanco-Ulate, B., Powell, A.L.T., Harding, S.E., Craigon, J., Rose, J.K.C., Fich, E., Sun, L., Domozych, D.S., Fraser, P.D., Tucker, G.A., Grierson, D. and Seymour, G.B. (2016) Genetic improvement of tomato by selective control of fruit softening, *Nature Biotechnology*, 34, 950–952 (doi: 10.1038/nbt.3602) http://www.nature.com/nbt/journal/vaop/ncurrent/full/nbt.3602.html

Pectic Polysaccharides and Bioactivity

3

3.1 STRUCTURAL ASPECTS OF PECTINS

We now consider as a case study one class of structural polysaccharides - the pectins, and their bioactivity, which offer potential biotechnological/biomedical exploitation. As we have seen in the previous chapter, pectic polysaccharides are major constituents of the plant cell wall, and they are the most complex class of polysaccharides. Albersheim et al. (1996) have suggested that the same set of pectic subunits is present in the cell walls of all higher plants. These subunits include homogalacturonan (HG), rhamnogalacturonan I (RG-I) and substituted galacturonans like rhamnogalacturonan II (RG-II) and xylogalacturonan (XGA). Similar to other plant polysaccharides, pectins are polymolecular and polydisperse, exhibiting significant heterogeneity within their botanical origin, with respect to both chemical structure and molecular weight (Willats et al., 2001; Perez et al., 2003).

HG consists of linear chains of 1→4 linked α-D-galactopyranosyluronic acid (GalpA) residues, and comprises 65% of pectin. Some of the carboxyl groups are methyl esterified, and depending on the plant source the GalpA residues may also be O-acetylated in position 2 or 3. The ester substitution may be random or present as blocks of esterified or unesterified regions. Many of the properties and biological functions of HG are believed to be determined by ionic interactions (O'Neill and York, 2003; Mohnen, 2008).

RG-I contains a backbone of alternating 1→2 linked α-L-rhamnopyranosyl (Rhap) and 1→4 linked GalpA residues, and represents 20%–35% of pectin. The backbone GalpA residues in many RG-I are O-acetylated in position 2 and/or 3, but there is no evidence that the GalpA residues are methyl esterified. Newer studies show that there can be monosaccharides attached via position 3 of GalA units within the RG-I structure (Austarheim et al., 2012a), and it has been reported that a single β-D-glucuronosyl acid (GlcA) residue is attached to GalpA in sugar beet RG-I (O'Neill and York, 2003; Mohnen, 2008). Between 20% and 80% of the Rhap residues in the backbone are, depending on the plant source and the method of isolation, most often substituted in position 4 with neutral and acidic oligo- and polysaccharides. Recent studies have revealed that position 3 also can be a substitution point (Austarheim et al., 2012a). The side chains attached to Rha in the RG-I backbone are predominantly composed of linear and branched β-D-galactopyranosyl (Galp) and α-L-arabinofuranosyl (Araf) residues.

Pectic 1→4 linked β-ᴅ-galactan with non-reducing terminal arabinose (Ara) substituted in position 3 of some of the Gal*p* units is known as type I arabinogalactan (AG-I). Type II arabinogalactans (AG-II) comprise a highly branched polymer with a backbone of 1→3 linked Gal*p* units, some of which are substituted in position 6 with side chains of 1→6 linked Gal*p* units. Ara*f*-units might be substituted through position 3 of the 1→6 linked galactan side chains, or found terminally linked to the 1→6 linked Gal*p* side chain (Huisman et al., 2001; Yamada and Kiyohara, 2007; Bonnin et al., 2014). As AG-II co-extract with pectin, and separation of these are difficult, it is suggested they are covalently linked to the backbone (Vincken et al., 2003), but AG-II may also occur in a complex family of proteoglycans known as arabinogalactan-proteins (AGPs) (Willats et al., 2001). Arabinans are also found as side chains on RG-I, consisting of a 1→5 linked Ara*f* backbone that can become branched by links through position 2 and 3. The oligosaccharide side chains in RG-I from some plants may be esterified with phenolic acids (e.g. ferulic acid), and/or occasionally terminated by α-ʟ-fucosyl (Fuc), GlcA or 4-*O*-methyl-GlcA (Willats et al., 2001).

The relative proportions of the different side chains differ depending on the plant source, and their distribution along the RG-I backbone may vary. The number of glycosyl residues in the side chains may range from a single glycosyl residue to more than 50. The highly branched nature of RG-I has made it known as the hairy region of pectin, in contrast to HG domains which are known as the smooth region (Willats et al., 2001).

RG-II consists of a HG backbone of around nine Gal*p*A residues, and is not structurally related to RG-I. It has a low molecular weight (5–10 kDa) and is a branched pectic domain containing 11 different glycosyl residues. These include the rare sugars apiose (Api), aceric acid (AceA), 2-*O*-Me Fuc, 2-*O*-Me xylose (Xyl), 3-deoxy-ᴅ-*manno*-2-octulosonic acid (Kdo) and 3-deoxy-ᴅ-*lyxo*-2-heptulosaric acid (Dha). RG-II is one of the most unusual polysaccharides yet identified in nature. It appears to be the only major pectic domain that does not have significant structural diversity or modulation of its fine structure, but is substituted by four heteropolymeric side chains of known and consistent lengths. This is in contrast to RG-I, which is highly variable both in its fine structure and its occurrence within cell walls (O'Neill and York, 2003).

Although the fine structure of the various pectic constituents is largely known, it is still unclear how these structural elements are combined into the macromolecular structure. Both RG-I and RG-II are thought to be covalently attached to HG domains as they are not separated from each other by size-exclusion chromatography (SEC), and they all contain 1→4 linked Gal*p*A residues in their backbones. However, direct evidence for these linkages is still lacking. Whether RG-I occurs on the same chain as RG-II and whether the domains are attached to HG at reducing or non-reducing ends is also still unknown. At present, there is no evidence for a direct linkage between RG-I and RG-II (Willats et al., 2001; Yapo, 2011). The most depicted model of the pectic constituents, the traditional model, is that the HG and RG-I form one continuous backbone (Perez et al., 2003; Yapo, 2011) (Figure 3.1). It is not known whether there are different types of RG-I regions containing just arabinans, galactans or arabinogalactans, or if one RG-I region contains all the identified types of side chains. Alternative macromolecular structures of pectin have been postulated where HG constituents are depicted as side chains of RG-I, the 'RG-I backbone' model (Figure 3.2a) (Vincken et al., 2003), and most recently a model where both the traditional and the 'RG-I backbone' models are accounted for (Figure 3.2b) (Yapo, 2011).

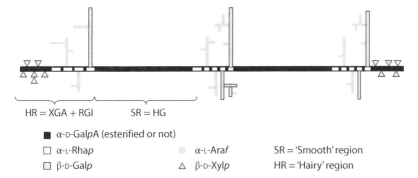

HR = XGA + RGI   SR = HG

- ■ α-D-GalpA (esterified or not)
- □ α-L-Rhap   α-L-Araf   SR = 'Smooth' region
- □ β-D-Galp   △ β-D-Xylp   HR = 'Hairy' region

FIGURE 3.1 The most commonly depicted model of the pectin-complex with alternating 'smooth' and 'hairy' regions. The model is assuming homogalacturonan (HG), rhamnogalacturonan I (RG-I) and xylogalacturonan (XGA) to be covalently linked, although this point is not firmly established. (From Yapo, B.M., *Carbohydrate Polymers*, 86, 373–385, 2011.)

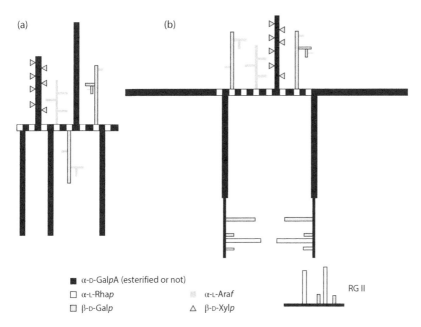

- ■ α-D-GalpA (esterified or not)
- □ α-L-Rhap   α-L-Araf
- □ β-D-Galp   △ β-D-Xylp

FIGURE 3.2 Alternative models of the pectin-complex. (a) A model after Vincken et al. (2003) showing HG constituents as side chains of RG-I. (b) A model after Yapo (2011) where the backbone of the pectin-complex is alternately composed of two linear HG elements and one RG-I core. The RG-I region is decorated with side chains of arabinans, galactans arabinogalactans and/or XGA, and may also include blocks of HG with RG-II linearly connected. (From Yapo, B.M., *Carbohydrate Polymers*, 86, 373–385, 2011.)

3.2 PECTIC POLYSACCHARIDES AS IMMUNOMODULATORS

Compounds that are capable of interacting with the immune system to upregulate or downregulate specific parts of the host response can be classified as immunomodulators. Polysaccharides such as β-glucans, inulin and pectins from various plants have attracted attention because of their effects on the immune system (Yamada and Kiyohara, 2007), in addition to their low toxicity and lack of harmful side effects (Austarheim et al., 2012b). Although the medical effect of the polysaccharides' influence on the human immune system remains to be determined, data suggest that they may participate in boosting the innate immune system in several ways such as through the complement system, by the influence on antigen-presenting cells like macrophages and dentritic cells (DCs), and possibly by interacting with other molecules and cells in the immune system. Dendritic cells are the most potent antigen-presenting cells of the immune system, orchestrating the initiation of the adaptive immune responses. Macrophages function as phagocytes and produce nitric oxide (NO), growth factors and cytokines upon activation. These effector functions are important in the immune response against an infection. The complement system plays an important role as a primary defense against bacterial and viral infections, and is a critical effector pathway in both adaptive and innate immunity, particularly in association with immunoglobulins. Due to the important physiological role of the complement immune system, complement modulation, either inhibition or stimulation, is related to various diseases and considered an interesting target for drug development (Alban et al., 2002). Complement modulation can be determined by a complement fixation method (Michaelsen et al., 2000).

There is a possibility that traditional medicines taken orally express their clinical effects through the intestinal immune system. Gut-associated lymphoid tissues (GALTs) exist along the intestinal wall and play important roles in host defense such as IgA production. Lymphocytes in Peyer's patches of GALT circulate through mesenteric lymph nodes in the intestine, and reach the systemic circulation to deliver into peripheral lymph nodes, spleen, and other lymphoid tissues, and hence the intestinal immune system may influence the systemic immune system as well as the mucosal immune system (Yu et al., 2001). Pharmacokinetic studies performed suggest that parts of the pectic polysaccharides are absorbed into the body through the intestine, and that they can be taken up via the Peyer's patches and then transported to systemic circulation through mesenteric lymph nodes and thoracic duct (Yamada and Kiyohara, 2007). A recent study has shown that galactans and arabinogalactans from modified citrus pectin might be absorbed via the paracellular pore (Courts, 2013).

Polysaccharides showing immunomodulating activities are exhibited in a wide range of glycans, extending from homopolymers to highly complex heteropolymers. Immunomodulating polysaccharides from the cell wall of medicinal plants frequently contain pectic polysaccharides with arabinogalactan side chains or pure arabinogalactans (Yamada and Kiyohara, 1999; Paulsen and Barsett, 2005). As hot water extracts of plants is a common way to prepare traditional medicine, it was therefore relevant to study bioactivities of high-molecular-weight compounds in the Malian extracts presented in this chapter.

3.3 PECTIC POLYSACCHARIDES AND IMPORTANCE FOR DISEASES IN MALI

In both developing countries and in most industrialized countries, there is a strong tradition for the use of medicinal plants to treat different types of diseases. In Mali, West Africa, approximately 80% of the population relies almost entirely on traditional medicines for their primary health care. The traditional medicine is mainly based on plants, and most of these plants have never been investigated for their chemical composition or pharmacological properties. The provision of safe and effective therapies based on traditional medicine could become a critical tool to increase access to health care. It is therefore of importance to study these plants to substantiate the traditional medical knowledge, and to evaluate their benefits, risks and limitations.

The Division of Pharmacognosy, Department of Pharmaceutical Chemistry, School of Pharmacy, University of Oslo, has since 1996 collaborated with the Department of Traditional Medicine (DMT) in Mali. DMT is housed in the National Institute for Research in Public Health (INRSP), and is a collaborating centre of the World Health Organization (WHO) for research in traditional medicine. The main objective of the collaboration is to assure that traditional medicine is used complementary to conventional medicine in a rational way based on scientific observations and experiments, assuming that medicines can be developed or improved from local resources, in particular medicinal plants. The main activities of DMT are registration of traditional healers and their use of medicinal plants, in addition to research and development of improved traditional medicines (ITMs). These phytomedicines are being produced from local plants in their natural form or in the form of infusion in ointments and syrups, and are being standardized according to traditional administration regimes. So far DMT has developed nine ITMs, and seven of these are recognized as essential and effective medicines in Mali (Diallo and Paulsen, 2000).

Traditional healers and herbalists are the main informants for DMT in the development of ITMs, and form the basis of the ethnopharmacological studies. The surveys determine which plants undergo phytochemical, toxicological and biological studies. Ethnopharmacological studies have been performed in several areas in Mali (Diallo et al., 1999; 2002; Inngjerdingen et al., 2004; Gronhaug et al., 2008; Togola et al., 2008a), and have revealed plants used in the treatment of, among other ailments, wounds, infections, malaria and gastric ulcers. Traditional healers from different regions and different ethnical groups, using the same plant to treat the same disease, substantiate the use of these plants.

Infectious diseases, malaria, gastric ulcer, wounds of different origin and fungal diseases are common ailments in Mali and contribute to the high degree of mortality in the country. An effective immune response is necessary to recover from these diseases, and therefore it has been of interest to search for immunomodulating components in medicinal plants. The ability of pectic polysaccharides to boost certain immune responses may, in part, explain some of the beneficial effects of medicinal plants, as they are easily obtained by extracting the plant material at 50°C–100°C, and hot water extracts of Malian medicinal plants are commonly used preparations.

3.4 STRUCTURE AND BIOACTIVITY OF PECTIC POLYSACCHARIDES FROM MALIAN MEDICINAL PLANTS

3.4.1 *Biophytum petersianum*

Biophytum petersianum Klotzsch (syn. *Biophytum umbraculum* (L.) DC) (Oxalidaceae) is a slender annual herb with stems up to 25 cm long, growing throughout the tropical and subtropical regions of Asia, Africa and Madagascar (Burkhill, 1997). *B. petersianum* is used in the treatment of malaria, fever, wounds and different types of skin disorders in Malian traditional medicine (Diallo et al., 2002). A bioactive pectic polymer was obtained from a 100°C water extract of the aerial parts of *B. petersianum*, after purification by gel filtration and fractionation by ion-exchange chromatography. The polymer exhibits complement fixation properties, and was able to stimulate macrophages and dendritic cells. Macrophages were stimulated to produce NO and IL-6, while dendritic cells showed improved antigen-presenting characteristics (Inngjerdingen et al., 2006, 2008; Gronhaug et al., 2011). The native polymer consists of a highly methyl esterified linear HG region which seems to alternate with ramified regions comprising highly branched RGs (Figure 3.3).

Different subunits of the native polymer were recovered after treatment with *endo*-polygalacturonase. An RG-I-like structure was the predominant feature of the highest molecular weight fraction, with short side chains of both AG-I and AG-II. The RG-I region had more potent bioactivities compared to the native polymer, and the AG-I and AG-II side chains were shown to be important in the expression of bioactivities.

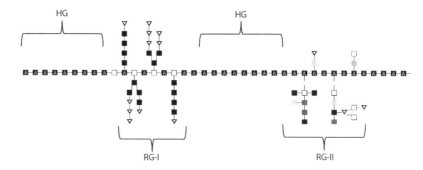

FIGURE 3.3 A proposed structure of the native bioactive pectic polysaccharide from *Biophytum petersianum*. Code as for Figures 3.1 and 3.2.

The absence of activation of B- and T-cell proliferation, as well as NK-cell activation, led to the conclusion that the pectic fractions from *B. petersianum* are not potent immunomodulators of lymphocytes (Inngjerdingen et al., 2008). This could mean that the fractions possibly recognize receptors that are preferentially expressed by antigen-presenting cells. The native polymer was further shown to protect against systemic *Streptococcus pneumoniae* infection in mice when administered intraperitoneally (i.p.) 3 h before challenge with the bacteria. The polymer induced the release of the cytokines IL-1α, IL-6, MCP-1, MIP-2, G-CSF and KC. The observed clinical effect is thought to occur via the native immune system (Inngjerdingen et al., 2013b).

3.4.2 *Cochlospermum tinctorium*

Cochlospermum tinctorium A. Rich. is of widespread occurrence in the savanna and scrubland throughout the drier parts of the West African Region, including Mali. In Malian traditional medicine, the roots of *C. tinctorium* are used, among other ailments, in the treatment of liver disorders and gastro-intestinal diseases, especially gastric ulcer and stomach ache (Nergard et al., 2005a). By anion-exchange chromatography, three acidic polysaccharide fractions were isolated from a 50°C water extract of the roots of *C. tinctorium* (Nergard et al., 2005c). The fractions, all containing AG-II-type structures (Figure 3.4), possessed complement fixation activities and led to a moderate induction of B-cell proliferation *in vitro* (Nergard et al., 2005a). Crude extracts and isolated polysaccharide fractions from *C. tinctorium* were, however, not able to stimulate mouse macrophages to release NO. This result may indicate that the fractions are not acting as general activators of human macrophages (Inngjerdingen et al., 2013a).

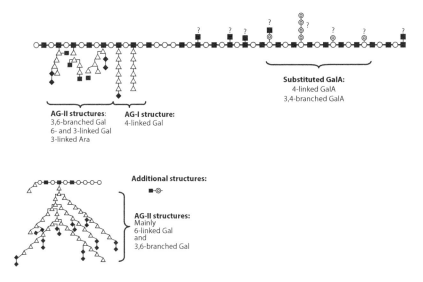

FIGURE 3.4   Proposed structures of the main pectic polysaccharides from *Cochlospermum tinctorium*.

A polysaccharide enriched extract, Ct50, was shown to significantly reduce the formation of gastric ulcer induced by HCl/ethanol in mice, in a dose-dependent manner (Nergard et al., 2005a). A crude 50°C water extract from the roots of *C. tinctorium*, and isolated pectic fractions rich in arabinogalactan structures, were shown to inhibit adhesion of *Helicobacter pylori* to human adherent gastric adenocarcinoma epithelial cells (AGS-cells) (Inngjerdingen et al., 2014).

3.4.3 *Cola cordifolia*

Cola cordifolia (Cav.) R.Br. (Malvaceae) is a large tree growing in the Savannah region between Senegal and Mali in Africa and the tree can be up to 25 m high. The trunk is rather short, branches grow fairly close to the ground and the foliage gives a dense crown providing good shadow for people and animals (Burkhill, 2000). All parts of the tree are used in traditional medicine, and in Mali especially, both bark and leaves are used for the treatment of different types of wound and stomach problems, pain, fever and diarrhea (Gronhaug et al., 2008; Togola et al., 2008c; Austarheim et al., 2012a). It is one of the most used plants in Mali for wound healing (Diallo et al., 2002). The most common administration is to suspend the powdered plant material in water and drink it or apply externally.

The polysaccharides of both bark and leaves have been studied in detail regarding both structure and biological activities, and the results of these studies will be discussed in the following.

The powdered bark from the trunk of *C. cordifolia* was, after pre-extraction with organic solvents, extracted with water of 50°C, and by purification by chromatography on two different kinds of anion exchange columns three different active polymers, CC1P1, CC1P2 and CC2, were obtained. Structural studies, including M_w determinations, were performed for the first two by a combination of monosaccharide and linkage analyses in addition to nuclear magnetic resonance (NMR) studies, while the structure of the latter is based only on monosaccharide and linkage analysis (Austarheim et al., 2012a, 2014).

CC1P1 is a new type of polymer with the structure [→2][α-D-Gal(1→3)] α-L-Rha(1→4)α-D-GalA(1→]$_{20}$. The backbone is similar to the RG-I backbone in pectins, but no smooth regions were found. The other special feature is that Gal is linked to position 3 of all Rha units via an α-linkage. Generally, Gal is linked via position 4 of Rha via a β-linkage (Figure 3.5) (Austarheim et al., 2012a). The polymer CC1P2 appeared to have the same backbone as CC1P1, but in addition to the terminal α-Gal units, α-4-OMe-GlcA and α-2-OMe-Gal were present as terminal units. A few substitutions on position 4 of Rha as well as on position 2 and 3 of GalA were observed, as was 1→4 linked Gal and terminal Ara (Figure 3.5). The polymer CC2 was not at all related to the structure of the two above-mentioned polymers, but had a structure more similar to general pectins containing AG-II-type structures, with a smaller amount of AG-I and arabinans as side chains on the RG-I backbone (Figure 3.5).

The very simple polymer CC1P1, and CC1P2, both had strong complement fixating activities compared with the standard that was used (PMII

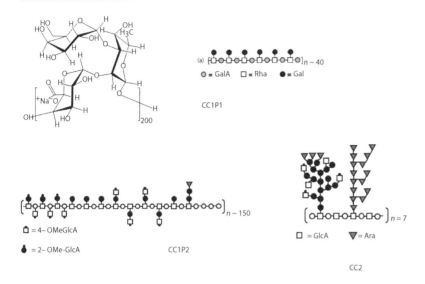

FIGURE 3.5   Proposed structures of the pectic polysaccharides studied from the bark of *Cola cordifolia*.

from *Plantago major*; Michaelsen et al., 2000), though CC1P2 had somewhat lower activity than the simple polymer CC1P1. Both were also tested for the ability to activate macrophages, and the interesting difference was that the simple polymer CC1P1 did not activate the macrophages while CC1P2 stimulated them to the same extent as the positive control (Austarheim et al., 2012a). The polymer CC1P1 had a higher activity in the complement assay than both the other polymers from the bark of *C. cordifolia*. The possible structural explanation for this higher activity may be that the two latter have acidic polymers attached on the outer part of the polymer, while CC1P1 has only the neutral Gal attached to the backbone. This explanation is in agreement with the theory that first was suggested by Yamada and Kiyohara (1999), saying that acidic units in the outer part of the molecule reduces the activity. CC1P1 has been further shown to protect against systemic *S. pneumoniae* infection in mice when administered i.p. (intraperitoneal injection) 3 h before challenge with the bacteria. The polymer induced the release of the cytokines IL-1α, IL-6, MCP-1, MIP-2, G-CSF and KC. The observed clinical effect is thought to occur via the native immune system (Inngjerdingen et al., 2013b).

The polysaccharides of the leaves have also been studied and the reason for doing this is that the bark of the tree is scraped off and used substantially as a wound healing remedy, and the removal of too much bark may ruin the tree. Thus, it was of interest to investigate the polysaccharides from the leaves to see if they had similar properties chemically and biologically to those bark polysaccharides. Leaf polysaccharides were studied both from the 50% ethanol/water and the 50°C water extracts. When purified in the same manner as the bark extracts, two polysaccharide fractions from each of the leaf extracts were obtained. They all had structural features similar to that of CC2 apart from one of the fractions from the ethanol/water extract (LCC50%A-P2). This one had even more acidic units in the outer part of the molecule being represented by higher amounts of 4-OMe-GlcA

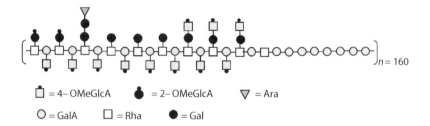

FIGURE 3.6   A proposed structure of LCC50%A-P2 from the leaves of *C. cordifolia.*

(Figure 3.6). Comparing the results of the complement fixation assay, the leaf polysaccharides all had somewhat lower activities than CC1P1 and CC1P2, but similar to that of CC2.

The polysaccharide-rich fractions both from the bark and the leaves were tested in a rodent model for anti-ulcer activity. Both the bark and leaf polysaccharides showed in a dose-dependent manner anti-ulcer activities which may explain the use of this medicinal tree against gastric ulcer (Austarheim et al., 2012b). The conclusion of the studies on the polysaccharides from both the bark and the leaves of *C. cordifolia* is that they all appear to have effects on the immune system. Although the simplest polymer had the highest effect in the complement assay, this had no effect on the stimulation of macrophages. Thus, from a biodiversity point of view, it could be better to encourage the local people to use the leaves instead of the bark as they would have the same total biological effect as using the bark.

3.4.4 *Combretum glutinosum*

Combretum glutinosum Perr. ex DC (Combretaceae) grows in the Sahel region. The tree can be up to 14 m high, but it may also grow like a bush. In West Africa, this tree is one of the most used medical remedies. The leaves are reported – used traditionally – as a diuretic, a cholagogue, against bronchitis, severe cough, cold, stomach spasms, gastric ulcer, diarrhea, dysmenorrhoea, amenorrhea, malaria, bilharzias, fever, otitis and for treatment of wounds (Neuwinger, 2000; Glaeserud et al., 2011). Previous ethnopharmacological studies performed in different regions of Mali on the use of the leaves of the tree *C. glutinosum* showed that the use against wounds and gastric ulcer were quite common (Diallo et al., 2002; Inngjerdingen et al., 2004). Information was given by traditional healers and herbalists. In Mali, the tree *C. glutinosum* can be found with two types of leaves, large and small, growing on different branches on the same tree. Some of the healers in the ethnopharmacological survey preferred to use the small leaves if they were available.

Pectic polysaccharides from small and large leaves of *C. glutinosum* were isolated and structurally characterized (Glaeserud et al., 2011). Defatted leaf preparations were extracted with 50°C water followed by 100°C water extraction. Acidic and high-molecular-weight fractions were isolated by anion-exchange chromatography and gel filtration. Small and large leaves were treated in exactly the same way giving two fractions, Cg50AP-S and Cg100AP-S, from small leaves, and two fractions, Cg50AP-L and Cg100AP-L, from large leaves. Their bioactivities were tested in the human complement fixation assay, as well as their ability to produce NO from macrophages and to elicit cytokine release from B-cells and dendritic cells. Both the fractions from the small leaves are more active in the complement fixation test than those from the large leaves. The Cg100AP-S fraction is the one with the highest activity. The acidic high-molecular-weight fraction isolated from the 50°C water extract of the small leaves, Cg50AP-S, was the only fraction giving macrophage activation.

Cg50AP-S and Cg50AP-L were almost equally potent inducers of production and secretion of the cytokines IL-6, IL-10 and TNF-α from dendritic cells, but not for IL-1α where Cg50AP-S induced a significantly higher amount than Cg50AP-L. Cytokine release from B cells was limited in response to both Cg50AP-S and Cg50AP-L, but there was a tendency for higher release of IL-10 after stimulation with Cg50AP-S compared to Cg50AP-L.

Structural investigation of the four polysaccharide fractions showed the presence of RG-I in all polysaccharides studied. AG-II was present in all four fractions, and this was confirmed by precipitation with Yariv reagent (Paulsen et al., 2014). Special differences between the polysaccharides isolated from the small leaves and the large leaves were the large amount of branch points on position 3 of the GalA units in the polysaccharides from small leaves (Figure 3.7) compared to the parallel fractions from the large leaves. The amount of branch points was fitting with the amount of terminal Xyl, and the authors concluded that the polymers contained regions of XGA. The somewhat higher bioactivities in the small leaves could be linked to the presence of the XGA region of these polymers.

3.4.5 *Entada africana*

Entada africana Guill. & Perr, Leguminosae, is a tree that is used in traditional medicine in several African countries, amongst those Mali. All plant parts are used against illnesses like fever that can be related to malaria (Doumbia, 1984; Bah, 1998); other ethnopharmacological studies

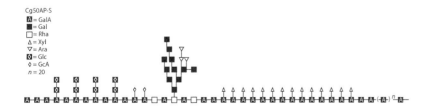

FIGURE 3.7   A proposed structure of the pectic polysaccharide present in the small leaves from *Combretum glutinosum*.

have shown that the tree, and especially the roots of it, is used as hepato-protecting and anti-inflammatory agents and for wound healing (Diallo et al., 2001). Decoction is the main preparation used, and due to this both chemical and biological studies were performed on the water-soluble compounds present in the root, and especially on the polysaccharides.

The polysaccharides were extracted with water at 50°C and 100°C successively from root material that previously had been defatted with organic solvents. After purification by anion-exchange chromatography followed by gel filtration, structural analyses and determination of activity in the complement fixation assay were performed. The EA50°C fraction consisted of an AG-II-type structure with high amounts of Rha and GlcA, while the acidic fractions present in the 100°C extract (EA100°C) differed in composition. The least acidic fraction, EA100°C fr. 1, consists of fragments with AG-II structures and xyloglucans. The latter with a 1→4 linked glucan backbone to which Xyl units are attached via position 6 of glucose (Glc). Araf is attached on position 4 on some of the Xyl units. Weak acid hydrolysis removed most of the Araf and revealed Xyl as terminal units. This polymer contained approximately 10% protein rich in hydroxyproline, serine and alanine. The other two fractions from the 100°C extract had a composition typical for pectins with a RG-I backbone. The side chains had typical AG-II structures, shown by the binding of the Yariv reagent (Diallo et al., 2001).

The polysaccharide EA50°C had a very low response in the complement fixation assay, while fractions from the EA100°C extract had significant effects in this bioassay. The most active fraction was EA100°C fr.1 having an ICH50 (concentration of test sample giving a 50% inhibition of lysis of red blood cells) value similar to that of the positive control PMII (Michaelsen et al., 2000), while the other two fractions from EA100°C had very low activities. Weak acid hydrolysis of all the fractions gave a loss of all effects in the complement fixation assay, indicating that for EA100°C fr.1 the terminal Araf units appeared to be important for the biological effect.

3.4.6 *Glinus oppositifolius*

Glinus oppositifolius (L.) Aug. DC. (Aizoaceae) is a slender spreading or ascending annual herb, growing on damp sandy sites. The plant occurs across West Africa from Senegal to South Nigeria, and is widely distributed in the tropics and subtropics generally (Burkhill, 1985). *G. oppositifolius* is used in Malian traditional medicine to treat wounds, inflammations, gastric ulcer, malaria and fever. By anion-exchange chromatography, two structurally different pectic polymers, GOA1 and GOA2, were isolated from a 50°C water extract of the aerial parts of *G. oppositifolius*. The polymers exhibit complement fixation activity, chemotactic properties towards macrophages, T cells and NK cells and are able to secrete IL-1β by macrophages. In addition, they were shown to give a marked increase of mRNA for IFN-γ in NK-cells (Inngjerdingen et al., 2005, 2007a,b). GOA1 consists of a small RG-I moiety to which complex side chains of both AG-I and AG-II are attached (Figure 3.8).

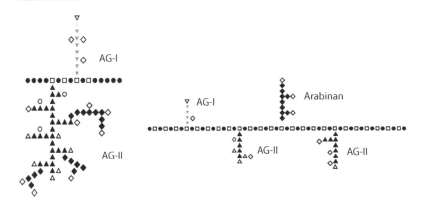

FIGURE 3.8   Tentative structures of two pectic polysaccharides from *Glinus oppositifolius.*

The arabinogalactan side chains were shown to be important for the bioactivity (Inngjerdingen et al., 2007a). GOA2 is rich in GalA and constitutes a typical pectic polymer (Figure 3.8). Treatment of GOA2 with endo-α-D-(1→4)-polygalacturonase led to the isolation of a potent bioactive fraction, GOA2-I. GOA2-I contains an RG-I-like structure with side chains of both AG-I and AG-II, and arabinans (Inngjerdingen et al., 2007a). The ramified region GOA2-I exhibited more potent bioactivities compared to its native polymer GOA2, suggesting that the bioactivity is expressed by the RG-I region (Inngjerdingen et al., 2007a). The immunomodulating properties of GOA1 and GOA2-I were further demonstrated by their abilities to proliferate B cells (Inngjerdingen et al., 2007a).

3.4.7 *Opilia celtidifolia*

Opilia celtidifolia (Guill. & Perr.) Endl. Ex Walp. (Opiliaceae) is a woody climber, heavily branched scrub or tree up to 10 m high, common in fringing forest and savanna. It is widespread in the region from Senegal to Nigeria (West Africa) and dispersed over the dried part of tropical Africa (Burkhill, 1997). *O. celtidifolia* is well known to the traditional healers in Mali as a remedy to cure several diseases, including all kinds of skin disorders and wounds, malaria, fever, pain and intestinal worms. The leaves of *O. celtidifolia* are the most frequent plant part used in traditional medicine, and a decoction is the most common preparation. A crude pectic polysaccharide extract was isolated from the leaves of *O. celtidifolia* by boiling water extraction of defatted leaf preparation (Sutovska et al., 2010). The polysaccharide extract showed significant biological effects on chemically induced cough reflex and reactivity of airways smooth muscle *in vitro* and *in vivo* conditions in a guinea pig test system. *In vitro* experiments confirmed a bronchodilatory effect of the *O. celtidifolia* extract, and revealed that the mechanism was partly accompanied by enhanced NO production.

Pectic-type polysaccharides were isolated by extraction of the defatted leaf preparation with 50°C water (Oc50) and 100°C water (Oc100) (Togola, et al., 2008b; Gronhaug et al., 2010). The Oc50 contained polysaccharides with the highest bioactivity as detected by the complement fixation assay, and was chosen for further studies. After gel-filtration and anion-exchange chromatography, two high molecular weight and acidic pectic polysaccharide fractions (Oc50A1 and Oc50A2) showed high activity in the complement system. These fractions were subjected to de-esterification and treatment with pectinase, and the most ramified region from each fraction (Oc50A1d1 and Oc50A2d1, respectively) showed even higher activity in the complement fixation assay. The ramified regions isolated had high amounts of Ara and Gal. The Oc50A1d1 contained 64% Ara and Gal totally, and Oc50A2d1 contained 52.2%. Both subfractions showed high contents of terminal, 1→3, 1→6 and 1→3,6 linked Gal (totally 83% and 62% of the Gal contents, respectively) and terminal and 1→5 linked Ara, and both precipitated with the Yariv reagent (Paulsen et al., 2014). Both subfractions contained 1→4 linked Gal in a somewhat smaller amounts and a relatively low amounts of 1→2 linked Rha and 1→4 linked GalA. The highly active subfractions seem to contain high amounts of AG-II, lower amounts of AG-I and a relatively short RG-I chain (Togola et al., 2008b).

The Oc50A1 fraction was further purified by gel filtration, first on a Sephacryl 400S column and then on a Superdex 200 column (Gronhaug et al., 2010). The purified high-molecular-weight fraction, Oc50A1.Ipur, was shown to give a potent production of the cytokines IL-1α, IL-6 and TNF-α from rat dendritic cells, and IL-10 from B cells. This suggests that the pectic polysaccharides may promote beneficial inflammation, since potent production of the pro-inflammatory cytokines IL-1α, IL-6 and TNF-α was observed. However, the release of the anti-inflammatory cytokine IL-10 from B cells suggests the pectins may also contribute to dampening or regulating the strength of the inflammatory processes (Gronhaug et al., 2010). The elution peak of Oc50A1.Ipur from the Superdex 200 gel filtration was divided into four fractions (-A, -B, -C, -D). Rechromatography of each subfraction showed that these four subfractions had different molecular weights, Oc50A1.IA having the highest M_w and Oc50A1.ID the lowest. The four subfractions showed large differences in complement fixation ability, and the -A fraction had the absolute highest activity. Fraction -A was the most potent inducer of NO, showed the highest increase in CD86 expression by dendritic cells and induced the highest B-cell proliferation activity of all the subfractions. The carbohydrate composition and the linkage analysis of the four subfractions showed that these fractions are similar, but not identical. They contained from 69.8% to 84.7% Ara and Gal, mostly presented as AG-II structures, and they have short RG-I chains (Figure 3.9). The reason for the great differences in bioactivity might be the differences in molecular size, shape/conformation or degree of polymerization (Gronhaug et al., 2010). The polymer Oc50A1.IA has further shown to protect against systemic *S. pneumoniae* infection in mice when administered i.p. 3 hours before challenge with the bacteria. The polymer induced the

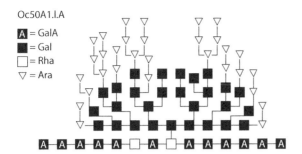

Oc50A1.I.A

A = GalA
■ = Gal
□ = Rha
▽ = Ara

FIGURE 3.9   A proposed structure of a pectic polysaccharide fraction present in *Opilia celtidifolia.*

release of the cytokines IL-1α, IL-6, MCP-1, MIP-2, G-CSF and KC. The observed clinical effect is thought to occur via the native immune system (Inngjerdingen et al., 2013b).

3.4.8 *Parkia biglobosa*

Parkia biglobosa (Jacq.) G. Don (Leguminosae), popularly called the African locust bean tree, can be up to 20 m high, and is one of the grain legumes (Anderson and De Pinto, 1985). In traditional medicine in Mali, the stem bark of *P. biglobosa* is used most frequently, followed by the leaves and seeds (Diallo et al., 2002; Gronhaug et al., 2008). Plant material, most often prepared as decoction, is used to cure a wide range of illnesses, most frequently wounds, pain and infections.

Pectic-type polysaccharides were isolated from defatted stem bark preparation by extraction with 50% ethanol (PBE), and 50°C and 100°C water (Zou et al., 2014). The water extracts were purified by ultrafiltration and dialysis giving high-molecular-weight fractions PB50 and PB100, respectively. After anion-exchange chromatography of PBE, PB50 and PB100, the fractions most active in the complement fixation assay were submitted to further fractionation by gel filtration. This gave seven active acidic subfractions. The two most active subfractions in the complement system were both from the 50% ethanol extract, and the same fractions (PBEII-I and PBEII-IV) showed statistically significant stimulating effects of macrophages based on NO release. All seven isolated fractions were pectic-type polymers. The two most active subfractions (PBEII-I and PBEII-IV) with molecular weights of 303.7 and 5.9 kDa, respectively, both contained terminal, 1→5 and 1→3,5 linked Ara, terminal and 1→2,4 linked Rha, terminal, 1→3, 1→6 and 1→3,6 linked Gal and 1→4 linked GalA, typical for pectins containing RG-I and AG-II structures. Previously, it has been reported that acidic polysaccharides with high molecular weights appear to be more active in the complement fixation assay than those with low molecular weights (Nergard et al., 2005a; Togola et al., 2008b; Gronhaug et al., 2010). The subfraction PBEII-IV with low molecular weight (5.9 kDa) may have side chains esterified with phenolic acids (Levigne et al., 2004), since 2.4% phenols were detected. This could have an influence on the biological activity detected.

3.4.9 *Syzygium guineense*

Syzygium guineense (Willd.) DC, Myrtaceae, is a flowering tree growing wild in sub-Sahara as well as in southern regions of Africa. In Mali, Senegal and Sierra Leone a decoction of its leaves are utilized in the traditional medicine against various ailments such as wound healing, ulcers, diarrhea, rheumatism and infections (Burkhill, 1997; Djoukeng et al., 2005).

Defatted leaves from *S. guineense* were extracted with 50°C water, desalted and fractionated on an anion-exchange chromatography column. The acidic fraction was divided into a pectic polysaccharide fraction, Sg50A1, and a pectic oligosaccharide fraction, Sg50A2, by gel filtration. The pectic polysaccharide fraction (M_w 36 kDa) contained an AG type II polysaccharide and the oligosaccharide fraction seemed to contain a mixture of pectic oligosaccharides with 10–15 units and a very short RG backbone with a relatively short AG type II side chain (Ghildyal et al., 2010). Both the polysaccharide and the oligosaccharide fraction showed higher complement fixation ability than the control polysaccharide, PMII, (Michaelsen et al., 2000), and Sg50A1 showed higher activity than Sg50A2. Determination of the ability of Sg50A1 and Sg50A2 to activate macrophages was assessed by measuring the production of NO. A dose-dependent release of NO was observed when macrophages were stimulated with Sg50A1 or with Sg50A2. The fractions caused an almost fourfold increase in NO-release at 100 µg/mL compared to the negative control. Sg50A1 and Sg50A2 further activated dendritic cells by up-regulating CD86 on rat bone marrow–derived dendritic cells, and induced B-cell proliferation. Sg50A1 and Sg50A2 were equally potent inducers of production and secretion of the cytokines IL-6, IL-10, and TNF-α from B cells and IL-1α, IL-6, IL-10 and TNF-α from dendritic cells. The analysis suggests that the pectic carbohydrate fractions stimulated the secretion of both pro-inflammatory and anti-inflammatory cytokines, which may contribute to dampening or regulating the strength of the inflammatory processes.

Sg50A1 and Sg50A2 differ in structure. The Sg50A1 is a high-molecular-weight AG type II polysaccharide and Sg50A2 is a mixture of oligosaccharides with typical pectin structures. Sg50A2 also contain typical linkages of AG type II, indicating that the presence of arabinogalactans is a determining factor for the bioactivity of both the polysaccharide and the oligosaccharides (Ghildyal et al., 2010).

3.4.10 *Terminalia macroptera*

Terminalia macroptera Guill. & Perr. (Combretaceae) is a tree that grows in West Africa. The root bark, stem bark and leaves of the tree are all used frequently in traditional African folk medicine. In Mali, *T. macroptera* is used against a variety of ailments; about 31 different indications have been mentioned by the traditional healers in ethnopharmacological studies, including wounds, hepatitis, diabetes, malaria, fever, cough and diarrhea as well as tuberculosis and skin diseases (Diallo et al., 2002;

Sanon et al., 2003; Pham et al., 2011). The stem bark and leaves are most commonly used against sores and wounds, pain, cough, tuberculosis and hepatitis. The roots are used against hepatitis, gonorrhea and various infectious diseases, including *H. pylori*-associated diseases (Silva et al., 1996, 1997, 2000, 2012; Pham et al., 2011).

The polysaccharides from three plant parts of *T. macroptera* have been studied, namely, root and stem bark and the leaves. The plant parts were extracted in three different ways: (1) conventional extraction with 50°C and 100°C water of plant material pre-extracted with organic solvents, (2) successive extraction with organic solvents and water at 50°C and 100°C in an Accelerated Solvent Extractor (ASE) and (3) extracted with boiling water the way the healers do, followed by separation into high- and low-molecular-weight compounds. The crude extracts and high-molecular-weight fractions were all further fractionated by ion-exchange chromatography and gel filtration (Zou et al., 2014, 2015). In principle the polysaccharides from each plant part were similar independent of the extraction process used, as were the bioactivities. Thus, the basic structures of the polysaccharides will be discussed in the following.

The polysaccharides are all of pectic nature. They contain typical HGs, with 1→4 linked GalA as the main chain. A few branch points in position 3 of GalA were present and these could be part of either RG-I or HG. RG-I had branches on position 4 of Rha, and the RG-I areas had different lengths depending on the parent polymer. The low ratio of Rha (including 1→2 linked and 1→2,4 linked Rha) to GalA (including 1→4 linked and 1→3,4 linked GalA) indicated that the backbone of most of the polysaccharide fractions consists of short RG-I structures, and longer HG regions (Zou et al., 2014; Zou et al., 2014; Zou et al., 2014; Zou et al., 2014; Zou et al., 2015). The polysaccharides from the trunk bark contain longer RG-I regions compared with the other purified fractions, since a higher ratio of 1→2 linked Rha to 1→4 linked GalA was found (Zou et al., 2014). The structural features found in the *T. macroptera* fractions have similarities with pectins that are composed of areas with hairy and smooth regions, and the different amount of linkages among fractions may lead to length variation of some of the structural regions (Vincken et al., 2003).

The presence of 1→3 linked Gal and 1→3,6 linked Gal concomitant with the precipitation with the Yariv reagent (Paulsen et al., 2014) showed the presence of AG-II structures in most of the polysaccharides isolated. The occurrence of AG-II, based on linkage analyses in one of the trunk bark fractions, TRBD-I-II, could not be detected by the Yariv-test although 1→3,6 linked Gal was present. This may be due to the fact that the chains of 1→3 linked Gal present in this polymer are too short (Paulsen et al., 2014). 1→4 linked Gal was only found in the fraction TRBD-I-II (from the trunk bark) which may indicate the presence of AG-I in this fraction only. Xylose was present in some of the fractions, mainly as 1→4 linked units, indicating that they may exist as xylans in the cell wall (Waldron and Faulds, 2007). GlcA is also present, terminally and as 1→4 linked units. Terminal GlcA might be directly linked to position 3 of 1→4 linked GalA in the RG-I backbone, or may also be a part of the AG-II side chains (Capek et al., 1987; Renard et al., 1999).

High amounts of phenolic compounds (30.8%) were found in some fractions, indicating that the side chains may be esterified with phenolic acids (Levigne et al., 2004).

One fraction from each plant part, 50WTRBH-II-I, 50WTSBH-II-I and 50WTLH-II-I, was chosen for enzymatic degradation by pectinase. Highly branched Gal units (1→3,4-linked and 1→3,6-linked) found in 50WTRBH-II-Ia indicated the presence of both AG-I and AG-II side chains in the hairy region obtained from the trunk polysaccharide. The analyses of the root bark native fraction and subfraction revealed that the native fraction consists of long HG regions and a shorter RG-I region with AG-I and AG-II side chains.

After pectinase treatment, three main fractions were obtained from the native fraction 50WTLH-II-I from the leaves. The highest M_w subfraction 50WTLH-II-Ia consists of an RG-I region with AG-I and AG-II side chains, and an RG-II region due to the presence of KDO. The analyses of the native fraction and subfractions suggest that the main feature of 50WTLH-II-I is 1→4 linked galacturonan, interrupted by RG-I and RG-II regions.

For the stem bark pectin, the main active site had a very low M_w (a small RG-I region, Figure 3.10). The analyses suggest that the main structural features of the stem bark pectin is a long HG region, with only a small RG-I region being highly active in the complement fixation assay (Zou et al., 2015).

3.4.11 *Trichilia emetica*

Trichilia emetica subsp. suberosa JJ de Wilde, Meliaceae, is a tree that can be up to 10 m high. The tree has its habitat mainly in tropical and subtropical regions, and is frequently found in Mali and other African countries, where it is an important medicinal tree (Burkhill, 1985). In Mali, the leaves of the tree are used primarily against old wounds, infested and cancerous, as well as sexual wounds. The water extract (decoction) is used for

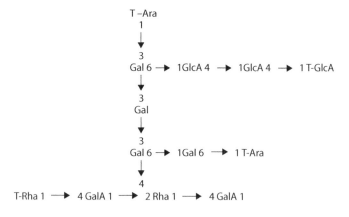

FIGURE 3.10 A proposed structure of the pectic fraction 50WTSBH-II-Ia from *T. macroptera*.

washing the wounds followed by application of powdered leaves (Diallo et al., 2002).

Dysmenorrhoea is also an ailment for which water extracts of *T. emetica* is used (Sanogo, 2011). Due to the traditional use against coughs, the antitussive activity was tested for the 100°C water extract of the leaves of *T. emetica*. The polysaccharide fraction showed significant cough-suppressive effect on chemically induced cough in a cat model and the results indicated effects related to bronchodilatory properties (Sutovska et al., 2009).

The water extracts (50°C and 100°C successively extracted) of leaves pre-extracted with organic solvents were shown to have effects in the complement fixation assay (Diallo et al., 2003), and due to this they were studied further from a structural and activity viewpoint. The crude extracts were separated by anion-exchange chromatography to give one acidic fraction from the 50°C water extract (Te50-1) and one neutral and four acidic ones fractions (Te100-1,2,3,4) from the 100°C water extract.

Te50-1 was the fraction with the highest effect in the complement fixation assay. This was mainly an arabinogalactan with small amounts of Rha and GalA present. The RG-I backbone is very small, but could accommodate the larger chains of AG-II attached to position 4 of the Rha units. Removal of Ara*f* units by weak acid hydrolysis revealed that those units primarily were linked both to positon 3 of 1→3,6 and 1→3 linked Gal to give increased amounts of 1→6 and terminal Gal units. Similar structural parts were also seen for the acidic polysaccharides extracted with 100°C water. It was interesting to note that the polysaccharide from the 50°C water extract had higher activity in the complement fixation assay than those extracted with water of 100°C. This may be due to differences in the orientation of the polymer, as well as molecular sizes that were different (Diallo et al., 2003).

3.4.12 *Vernonia kotschyana*

Baccharoides adoensis var. kotschyana (Sch. Bip. ex Walp.) M. A. Isawumi, G. El-Ghazaly and B. Nordenstam (Asteraceae), mainly described in the literature as *Vernonia kotschyana* Sch. Bip. ex Walp. or *Vernonia adoensis* Sch. Bip., is a shrub up to 2-m high growing on the savanna, from Senegal to Nigeria across Africa to Ethiopia (Burkhill, 1997). The name *V. kotschyana* will be used in this section as the new name has never been cited in the scientific literature. The roots of *V. kotschyana* are widely used in Mali in the treatment of gastrointestinal disorders and in wound healing (Nergard et al., 2004). A decoction of the powdered roots is recognized by the government in Mali as an ameliorated traditional medicine, Gastrosedal, for the treatment of gastritis and gastroduodenal ulcers, and this medicine is on the National List of Essential Drugs.

By anion-exchange chromatography, acidic polysaccharide fractions were isolated from 50°C and 100°C water extracts of the roots of *V. kotschyana* (Nergard et al., 2004). The acidic fractions containing AG-II-type structures, probably connected to a pectic backbone, showed both complement fixation and mitogenic activities. The acidic fractions

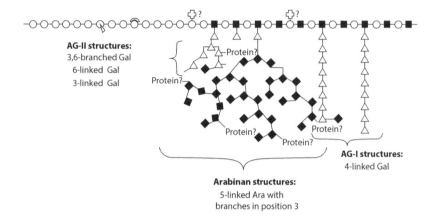

FIGURE 3.11 A proposed structure of the pectic fraction Vk2a from *Vernonia kotschyana.*

induced B-cell proliferation, but no effect was observed on T cells. A significant, but low activity was observed on macrophages (Nergard et al., 2004). The pectic fraction Vk100A2 obtained after ion-exchange chromatography of the 100°C water extract was further fractionated by SEC. A pectin (Vk100A2b) and a pectic arabinogalactan (Vk100A2a) were obtained. The pectin fraction, consisting mainly of 4-linked GalA, showed low complement fixation and no influence on the proliferation of B or T cells. The pectic arabinogalactans, Vk100A2a, however, showed potent dose-dependent complement fixation and a T-cell-independent induction of B-cell proliferation. Vk100A2a (Vk2a) consisted of a highly branched RG core with side chains rich in AG-I, AG-II and arabinan (Figure 3.11) (Nergard et al., 2005c), and the bioactive sites were suggested to be located both in the more peripheral parts of the molecule, but also in the inner core of the hairy region (Nergard et al., 2005b).

The polysaccharides were also shown to induce chemotaxis of human macrophages, T- and NK-cells, although donor variations were observed (Nergard et al., 2005c). Anti-ulcer activities of the 50°C and 100°C crude water extracts, consisting mainly of inulin, have been evaluated. The extracts were shown to inhibit the formation of gastric lesions in mice (100 mg kg^{-1}) (Austarheim et al., 2012c). A pectic arabinogalactan isolated from the roots of *V. kotschyana* was also shown to inhibit adhesion of *H. pylori* to human adherent gastric adenocarcinoma epithelial cells (AGS-cells) (Inngjerdingen et al., 2014).

3.5 DISCUSSION AND SUMMARY

The important structural requirements for the observed immunomodulating activities presented in this chapter seem to be branched structures involving arabinogalactan regions and an RG backbone (Figure 3.12). A certain three-dimensional configuration may also be important for the activities. More than one binding site in each molecule giving an overall shape

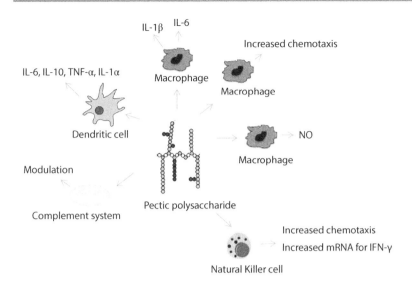

FIGURE 3.12   A summary of immunomodulating effects of pectic polysaccharides from Malian medicinal plants.

and surface may be important for the binding to factors involved in the bioactivities. For polysaccharides in vaccines (bacterial capsular polysaccharides), as we will consider in Chapter 6, it has been shown that immunogenicity is related to their molecular size and that in general the larger the molecule the more immunogenic it is likely to be. However, there is no absolute size above which a substance will be immunogenic. A correlation between immunogenicity and molecular weight of plant polysaccharides may therefore also exist. Conformational epitopes are suggested to be better expressed in larger saccharides, and larger molecules may also contain more than one epitope which may increase the immunogenicity of the substance (Paoletti et al., 1992; von Hunolstein et al., 2003).

Macrophages have for a long time been considered as one of the main target cells for polysaccharide interaction, and several compounds have been shown to modulate their cytokine production. Regarding pectins, it has been suggested that the β-D-(1→3,6)-galactan chains of AG-II play an important role in the immunomodulation of macrophages, and also the β-D-(1→4)-galactan chains of AG-I (Gronhaug et al., 2011). The importance of Ara for macrophage-stimulating activity has been shown for pectins from elderberries. A loss of Ara-units, and by this the 1→3,6 Gal branch point, by weak acid hydrolysis, led to a reduction in NO-release from macrophages (Ho et al., 2015). On the other hand, a removal of small amounts of Ara-units by exo-α-L-arabinofuranosidase in the RG-I side chains of a pectin from *B. petersianum* led to a significant increase in IL-6 secretion from macrophages (Gronhaug et al., 2011). AG-II containing polysaccharides isolated from *Artemisia tripartita* and *Juniperus scopolorum* have shown to induce the production of IL-6 and MCP-1 from murine macrophages (Schepetkin et al., 2005; Xie et al., 2008), while an AG-II fraction isolated from peach pulp has been reported to induce peritoneal macrophage activation (Matsumoto et al., 2008; Simas-Tosin et al., 2012).

Macrophage activation by plant polysaccharides is thought to be mediated primarily through the recognition of polysaccharide polymers by specific receptors. Toll-like receptor 4 (TLR4) has been identified as the main pattern recognition receptor expressed by macrophages and mediates macrophage activation by transmitting a variety of extracellular signals, resulting in the secretion of cytokines such as NO and TNF-α. An acidic polysaccharide containing GalA, Rha, Gal and Ara isolated from the roots of *Angelica sinensis* (Oliv.) Diels has earlier been reported to promote NO production by upregulating the expression of TLR4 on macrophages (Yang et al., 2007). The ability of the pectic polysaccharides from the Malian medicinal plants to stimulate TLRs is, however, unknown.

The majority of the plant polysaccharides reported to modulate the complement system are pectic polymers containing AG-II structural units with 1→3,6 branched or 6-linked Gal residues. The importance of Ara-units for the activity has been reported for arabinogalactans isolated from, amongst others, elderberries (Ho et al., 2015) and *E. africana* (Diallo et al., 2001). A reduction in complement fixation activity was observed after loss of Ara-units and the 1→3,6-Gal branch point by weak acid hydrolysis of active pectins. On the other hand, a removal of Ara-units in the outer parts of the RG-I side chains by enzymatic digestion did not alter the complement fixation activity for pectins isolated from the Malian medicinal plants *G. oppositifolius* and *V. kotschyana* (Nergard et al., 2005b,c; Inngjerdingen et al., 2007b). An explanation for this might be that different linkages are broken down after weak acid hydrolysis compared to enzymatic degradation. However, not all polysaccharides containing an AG-II structure exhibit effects on the complement system (Yamada and Kiyohara, 1999; Alban et al., 2002), and other plant polysaccharides such as arabinans and glucuronoarabinoxylan have shown complement modulation (Yamada and Kiyohara, 1999; Alban et al., 2002).

The presence of immunomodulating polysaccharides in Malian medicinal plants can be related to the medical effects. However, the putative immunomodulatory applications of the pectic polysaccharides studied here will need further investigation. An adjuvant effect with other already existing chemical alternatives could be a possibility. Degrading the pectic polymers further by different enzymes and assessing the influence of the enzymatic treatment on the bioactivities may shed light on to the minimal structure of the polysaccharide responsible for an optimal inducing effect on different immune cells.

FURTHER READING

Liu, J., Willfor, S. and Xu, C. (2015) A review of bioactive plant polysaccharides: Biological activities, functionalization, and biomedical applications. *Bioactive Carbohydrates and Dietary Fibre*, 5, 31–61.

Ramawat, K. G., Mérillon, J.M. (Eds.) (2015) *Polysaccharides. Bioactivity and Biotechnology*, Berlin: Springer Verlag.

SPECIFIC REFERENCES

Alban, S., Classen, B., Brunner, G. and Blaschek, W. (2002) Differentiation between the complement modulating effects of an arabinogalactan-protein from *Echinacea purpurea* and heparin, *Planta Medica*, 68, 1118–1124.

Albersheim, P., Darvill, A.G., O'Neill, M.A., Schols, H.A. and Voragen A.G.J. (1996) An hypothesis, the same six polysccharides are components of the primary cell walls of all higher plants, in Visser, J. and Voragen, A.G.J. (eds) *Pectins and Pectinases*, pp. 47–53, Amsterdam, the Netherlands: Elsevier.

Anderson, D.M.W. and De Pinto, G.L. (1985) Gum polysaccharides from three *Parkia* species, *Phytochemstry*, 24, 77–79.

Austarheim, I., Christensen, B.E., Aas, H.T.N., Thole, C., Diallo, D. and Paulsen, B.S. (2014) Chemical characterization and complement fixation of pectins from *Cola cordifolia* leaves, *Carbohydrate Polymers*, 102, 472–480.

Austarheim, I., Christensen, B.E., Hegna, I.K., Petersen, B.O., Duus, J.O., Bye, R., Michaelsen, T.E., Diallo, D., Inngjerdingen, M. and Paulsen, B.S. (2012a) Chemical and biological characterization of pectin-like polysaccharides from the bark of the Malian medicinal tree, *Cola cordifolia*, *Carbohydrate Polymers*, 89, 259–268.

Austarheim, I., Mahamane, H., Sanogo, R., Togola, A., Khaledabadi, M., Vestrheim, A.C., Inngjerdingen, K.T., Michaelsen, T.E., Diallo, D. and Paulsen, B.S. (2012b) Anti-ulcer polysaccharides from *Cola cordifolia* bark and leaves, *Journal of Ethnopharmacology*, 143, 221–227.

Austarheim, I., Nergard, C.S., Sanogo, R., Diallo, D. and Paulsen, B.S. (2012c) Inulin-rich fractions from *Vernonia kotschyana* roots have anti-ulcer activity, *Journal of Ethnopharmacology*, 144, 82–85.

Bah, S. (1998) Sensibilité d'Anopheles gambiae aux insecticides organiques de synthèse et á divers extraits de plantes médicinales du Mali, M-Pharm Dissertation, University of Bamako, Mali.

Bonnin, E., Garnier, C. and Ralet, M.C. (2014) Pectin-modifying enzymes and pectin-derived materials, applications and impacts, *Applied Microbiology and Biotechnology*, 98, 519–532.

Burkhill, H.M. (1985) *The Useful Plants of West Tropical Africa*, London: Royal Botanical Gardens at Kew.

Burkhill, H.M. (1997) *The Useful Plants of West Tropical Africa*, London: Royal Botanical Gardens at Kew.

Burkhill, H.M. (2000) *The Useful Plants of West Tropical Africa*, London: Royal Botanical Gardens at Kew.

Capek, P., Rosik, J., Kardosova, A. and Toman, R. (1987) Polysacchariddes from the roots of the marshmellow (*Althea officinalis* var. Robusta)—Structural features of an acidic polyscacharide, *Carbohydrate Research*, 164, 443–452.

Courts, F.L. (2013) Profiling of modifoed citrus pectin oligosaccharide transport across Caco-2-cell monolayers, *Pharma Nutrition*, 1, 22–31.

Diallo, D. and Paulsen, B.S. (2000) *Pharmeceutical Research and Traditional Parctitioners in Mali, Experiences With Benefit Sharing*, Oslo: Spartacus.

Diallo, D., Hveem, B., Mahmound, M.A., Berge, G., Paulsen, B.S. and Maiga, A. (1999) An ethnobotanical survey of herbal drugs of Gourma district, Mali, *Pharmaceutical Biology*, 37, 80–91.

Diallo, D., Paulsen, B.S., Liljeback, T.H.A. and Michaelsen, T.E. (2001) Polysaccharides from the roots of *Entada africana* Guill. et Perr., Mimosaceae, with complement fixing activity, *Journal of Ethnopharmacology*, 74, 159–171.

Diallo, D., Paulsen, B.S., Liljeback, T.H.A. and Michaelsen, T.E. (2003) The Malian medicinal plant *Trichilia emetica*; studies on polysaccharides with complement fixing ability, *Journal of Ethnopharmacology*, 84, 279–287.

Diallo, D., Sogn, C., Samake, F.B., Paulsen, B.S., Michaelsen, T.E. and Keita, A. (2002) Wound healing plants in Mali, the Bamako region. An ethnobotanical survey and complement fixation of water extracts from selected plants, *Pharmaceutical Biology*, 40, 117–128.

Djoukeng, J.D., Abou-Mansour, E., Tabacchi, R., Tapondjou, A.L., Bouda, H. and Lontsi, D. (2005) Antibacterial triterpenes from *Syzygium guineense* (Myrtaceae), *Journal of Ethnopharmacology*, 101, 283–286.

Doumbia, B. (1984) Contribution à l'étude des plantes médicinales utilisées dans le traitement du paludisme au Mali, *Memoire, Biologie*, Bamako: Ensup.

Ghildyal, P., Gronhaug, T.E., Skogsrud, R.A.M., Rolstad, B., Diallo, D., Michaelsen, T.E., Inngerdingen, M. and Paulsen, B.S. (2010) Chemical composition and immunological activities of polysaccharides isolated from the Malian medicinal plant *Syzygium guineense*, *Journal of Pharmacognosy and Phytotherapy*, 2, 76–85.

Glaeserud, S., Gronhaug, T.E., Michaelsen, T.E., Inngjerdingen, M., Barsett, H., Diallo, D. and Paulsen, B.S. (2011) Immunomodulating polysaccharides from leaves of the Malian medicinal tree *Combretum glutinosum*; structural differences between small and large leaves can substantiate the preference for small leaves by some healers, *Journal of Medicinal Plants Research*, 5, 2781–2790.

Gronhaug, T.E., Ghildyal, P., Barsett, H., Michaelsen, T.E., Morris, G., Diallo, D., Inngjerdingen, M. and Paulsen, B.S. (2010) Bioactive arabinogalactans from the leaves of *Opilia celtidifolia* Endl. ex Walp. (Opiliaceae), *Glycobiology*, 20, 1654–1664.

Gronhaug, T.E., Glaeserud, S., Skogsrud, M., Ballo, N., Bah, S., Diallo, D. and Paulsen, B.S. (2008) Ethnopharmacological survey of six medicinal plants from Mali, West-Africa, *Journal of Ethnobiology and Ethnomedicine*, 4, 26.

Gronhaug, T.E., Kiyohara, H., Sveaass, A., Diallo, D., Yamada, H. and Paulsen, B.S. (2011) Beta-D-(1→4)-galactan-containing side chains in RG-I regions of pectic polysaccharides from *Biophytum petersianum* Klotzsch contribute to expression of immunomodulating activity against intestinal Peyer's patch cells and macrophages, *Phytochemistry*, 72, 2139–2147.

Ho, G.T.T., Ahmed, A., Zou, Y.F., Aslaksen, T., Wangensteen, H. and Barsett, H. (2015) Structure-activity relationship of immunomodulating pectins from elderberries, *Carbohydrate Polymers*, 125, 314–322.

Huisman, M.M.H., Fransen, C.T.M., Kamerling, J.P., Vliegenthart, J.F.G., Schols, H.A. and Voragen, A.G.J. (2001) The CDTA-soluble pectic substances from soybean meal are composed of rhamnogalacturonan and xylogalacturonan but not homogalacturonan, *Biopolymers*, 58, 279–294.

Inngjerdingen, K.T., Ballo, N., Zhang, B.Z., Malterud, K.E., Michaelsen, T.E., Diallo, D. and Paulsen, B.S. (2013a) A comparison of bioactive aqueous extracts and polysaccharide fractions from roots of wild and cultivated *Cochlospermum tinctorium* A. Rich, *Phytochemistry*, 93, 136–143.

Inngjerdingen, K.T., Coulibaly, A., Diallo, D., Michaelsen, T.E. and Paulsen, B.S. (2006) A complement fixing polysaccharide from *Biophytum petersianum* Klotzsch, a medicinal plant from Mali, West Africa, *Biomacromolecules*, 7, 48–53.

Inngjerdingen, K.T., Debes, S.C., Inngjerdingen, M., Hokputsa, S., Harding, S.E., Rolstad, B., Michaelsen, T.E., Diallo, D. and Paulsen, B.S. (2005) Bioactive pectic polysaccharides from *Glinus oppositifolius* (L.) Aug. DC., a Malian medicinal plant, isolation and partial characterization, *Journal of Ethnopharmacology*, 101, 204–214.

Inngjerdingen, M., Inngjerdingen, K.T., Patel, T.R., Allen, S., Chen, X.Y., Rolstad, B., Morris, G.A., et al. (2008) *Pectic polysaccharides from Biophytum petersianum* Klotzsch, and their activation of macrophages and dendritic cells, *Glycobiology*, 18, 1074–1084.

Inngjerdingen, K.T., Kiyohara, H., Matsumoto, T., Petersen, D., Michaelsen, T.E., Diallo, D., Inngjerdingen, M., Yamada, H. and Paulsen, B.S. (2007a) An immunomodulating pectic polymer from *Glinus oppositifolius*, *Phytochemistry*, 68, 1046–1058.

Inngjerdingen, K.T., Langerud, B.K., Rasmussen, H., Olsen, T.K., Austarheim, I., Gronhaug, T.E., Aaberge, I.S., Diallo, D., Paulsen, B.S. and Michaelsen, T.E. (2013b) Pectic polysaccharides isolated from Malian medicinal plants protect against *Streptococcus pneumoniae* in a mouse pneumococcal infection model, *Scandinavian Journal of Immunology*, 77, 372–388.

Inngjerdingen, K., Nergard, C.S., Diallo, D., Mounkoro, P.P. and Paulsen, B.S. (2004) An ethnopharmacological survey of plants used for wound healing in Dogonland, Mali, West Africa, *Journal of Ethnopharmacology*, 92, 233–244.

Inngjerdingen, K.T., Patel, T.R., Chen, X.Y., Kenne, L., Allen, S., Morris, G.A., Harding, S.E., et al. (2007b) Immunological and structural properties of a pectic polymer from *Glinus oppositifolius*, *Glycobiology*, 17, 1299–1310.

Inngjerdingen, K.T., Thöle, C., Diallo, D., Paulsen, B.S. and Hensel, A. (2014) Inhibition of *Helicobacter pylori* adhesion to human gastric adenocarcinoma epithelial cells by aqueous extracts and pectic polysaccharides from the roots of *Cochlospermum tinctorium* A. Rich. and *Vernonia kotschyana*, Sch Bip. ex Walp, *Fitoterapia*, 95, 127–132.

Levigne, S., Ralet, M.C., Quemener, B. and Thibault, J.F. (2004) Isolation of diferulic bridges ester-linked to arabinan in sugar beet cell walls, *Carbohydrate Research*, 339, 2315–2319.

Matsumoto, T., Moriya, M., Sakurai, M.H., Kiyohara, H., Tabuchi, Y. and Yamada, H. (2008) Stimulatory effect of a pectic polysaccharide from a medicinal herb, the roots of *Bupleurum falcatum* L., on G-CSF secretion from intestinal epithelial cells, *International Immunopharmacology*, 8, 581–588.

Michaelsen, T.E., Gilje, A., Samuelsen, A.B., Hogasen, K., and Paulsen, B.S. (2000) Interaction between human complement and a pectin type polysaccharide fraction, PMII, from the leaves of *Plantago major* L, *Scandinavian Journal of Immunology*, 52, 483–490.

Mohnen, D. (2008) Pectin structure and biosynthesis, *Current Opinion in Plant Biology*, 11, 266–277.

Nergard, C.S., Diallo, D., Inngjerdingen, K., Michaelsen, T.E., Matsumoto, T., Kiyohara, H., Yamada, H. and Paulsen, B.S. (2005) Medicinal use of *Cochlospermum tinctorium* in Mali: Anti-ulcer-, radical scavenging- and immunomodulating activities of polymers in the aqueous extract of the roots, *Journal of Ethnopharmacology*, 96, 255–269.

Nergard, C.S., Diallo, D., Michaelsen, T.E., Malterud, K.E., Kiyohara, H., Matsumoto, T., Yamada, H. and Paulsen, B.S. (2004a) Isolation, partial characterisation and immunomodulating activities of polysaccharides from *Vernonia kotschyana* Sch Bip. ex Walp, *Journal of Ethnopharmacology*, 91, 141–152.

Nergard, C.S., Kiyohara, H., Reynolds, J.C., Thomas-Oates, J.E., Matsumoto, T., Yamada, H., Michaelsen, T.E., Diallo, D. and Paulsen, B.S. (2005b). Structure-immunomodulating activity relationships of a pectic arabinogalactan from *Vernonia kotschyana* Sch Bip. ex Walp, *Carbohydrate Research*, 340, 1789–1801.

Nergard, C.S., Matsumoto, T., Inngjerdingen, M., Inngjerdingen, K., Hokputsa, S., Harding, S.E., Michaelsen, T.E., et al. (2005c) Structural and immunological studies of a pectin and a pectic arabinogalactan from *Vernonia kotschyana* Sch Bip. ex Walp. (Asteraceae), *Carbohydrate Research*, 340, 115–130.

Neuwinger, H.D. (2000) *African Traditional Medicine*, Stuttgart: Medpharm Scientific Publishers.

O'Neill, M.A. and York, W.S. (2003) *The Composition and Structure of Plant Primary Cell Walls*, Oxford: Blackwell Publishing.

Paoletti, L.C., Kasper, D.L., Michon, F., Difabio, J., Jennings, H.J., Tosteson, T.D. and Wessels, M.R. (1992) Effects of chain-length on the immunogenicity in rabbits of group-b streptococcus type-III oligosaccharide-tetanus toxoid conjugates, *Journal of Clinical Investigation*, 89, 203–209.

Paulsen, B.S. and Barsett, H. (2005) Bioactive pectic polysaccharides, in Heinze, T (ed) *Polysaccharides 1, Structure, Characterization and Use*, Vol. 186, pp. 69–101, Berlin: Springer.

Paulsen, B.S., Craik, D.J., Dunstan, D.E., Stone, B.A. and Bacic, A. (2014) The Yariv reagent, Behaviour in different solvents and interaction with a gum arabic arabinogalactanprotein, *Carbohydrate Polymers*, 106, 460–468.

Perez, S., Rodriguez-Carvajal, M.A. and Doco, T. (2003) A complex plant cell wall polysaccharide, rhamnogalacturonan II, a structure in quest of a function, *Biochimie*, 85, 109–121.

Pham, A.T., Dvergsnes, C., Togola, A., Wangensteen, H., Diallo, D., Paulsen, B.S. and Malterud, K.E. (2011) *Terminalia macroptera*, its current medicinal use and future perspectives, *Journal of Ethnopharmacology*, 137, 1486–1491.

Renard, C., Crepeau, M.J. and Thibault, J.F. (1999) Glucuronic acid directly linked to galacturonic acid in the rhamnogalacturonan backbone of beet pectins, *European Journal of Biochemistry*, 266, 566–574.

Sanogo, R. (2011). Medicinal plants traditionally used in Mali for dysmenorrhoea, *African Journal of Traditional Complementary and Alternative Medicines*, 8, 90–96.

Sanon, S., Ollivier, E., Azas, N., Mahiou, V., Gasquet, M., Ouattara, C.T., Nebie, I., et el. (2003) Ethnobotanical survey and in vitro antiplasmodial activity of plants used in traditional medicine in Burkina Faso, *Journal of Ethnopharmacology*, 86, 143–147.

Schepetkin, I.A., Faulkner, C.L., Nelson-Overton, L.K., Wiley, J.A. and Quinn, M.T. (2005) Macrophage immunomodulatory activity of polysaccharides isolated from *Juniperus scopolorum*, *International Immunopharmacology*, 5, 1783–1799.

Silva, O., Duarte, A., Cabrita, J., Pimentel, M. Diniz, A. and Gomes, E. (1996) Antimicrobial activity of Guinea-Bissau traditional remedies, *Journal of Ethnopharmacology*, 50, 55–59.

Silva, O., Duarte, A., Pimentel, M., Viegas, S., Barroso, H., Machado, J., Pires, I., Cabrita, J. and Gomes, E. (1997) Antimicrobial activity of *Terminalia macroptera* root, *Journal of Ethnopharmacology*, 57, 203–207.

Silva, O., Gomes, E.T., Wolfender, J.L., Marston, A. and Hostettmann, K. (2000) Application of high performance liquid chromatography coupled with ultraviolet spectroscopy and electrospray mass spectrometry to the characterisation of ellagitannins from *Terminalia macroptera* roots, *Pharmaceutical Research*, 17, 1396–1401.

Silva, O., Viegas, S., de Mello-Sampayo, C., Costa, M.J.P., Serrano, R., Cabrita, J. and Gomes, E.T. (2012) Anti-*Helicobacter pylori* activity of *Terminalia macroptera* root, *Fitoterapia*, 83, 872–876.

Simas-Tosin, F.F., Abud, A.P.R., De Oliveira, C., Gorin, P.A.J., Sassaki, G.L., Bucchi, D.F. and Iacomini, M. (2012) Polysaccharides from peach pulp, structure and effects on mouse peritoneal macrophages, *Food Chemistry*, 134, 2257–2260.

Sutovska, M., Franova, S., Priseznakova, L., Nosal'ova, G., Togola, A., Diallo, D., Paulsen, B.S. and Capek, P. (2009) Antitussive activity of polysaccharides isolated from the Malian medicinal plants, *International Journal of Biological Macromolecules*, 44, 236–239.

Sutovska, M., Franova, S., Sadlonova, V., Gronhaug, T.E., Diallo, D., Paulsen, B.S. and Capek, P. (2010) The relationship between dose-dependent antitussive and bronchodilatory effects of *Opilia celtidifolia* polysaccharide and nitric oxide in guinea pigs, *International Journal of Biological Macromolecules*, 47, 508–513.

Togola, A., Austarheim, I., Theis, A., Diallo, D. and Paulsen, B.S. (2008a) Ethnopharmacological uses of *Erythrina senegalensis*, a comparison of three areas in Mali, and a link between traditional knowledge and modern biological sciene, *Journal of Ethnobiology and Ethnomedicine*, 4, 6–14.

Togola, A., Inngjerdingen, M., Diallo, D., Barsett, H., Rolstad, B., Michaelsen, T.E. and Paulsen, B.S. (2008b) Polysaccharides with complement fixing and macrophage stimulation activity from *Opilia celtidifolia*, isolation and partial characterisation, *Journal of Ethnopharmacology*, 115, 423–431.

Togola, A., Naess, K.H., Diallo, D., Barsett, H., Michaelsen, T.E. and Paulsen, B.S. (2008c) A polysaccharide with 40% mono-O-methylated monosaccharides from the bark of *Cola cordifolia* (Sterculiaceae), a medicinal tree from Mali (West Africa), *Carbohydrate Polymers*, 73, 280–288.

Vincken, J.P., Schols, H.A., Oomen, R., McCann, M.C., Ulvskov, P., Voragen, A.G.J. and Visser, R.G.F. (2003) If homogalacturonan were a side chain of rhamnogalacturonan I: Implications for cell wall architecture, *Plant Physiology*, 132, 1781–1789.

von Hunolstein, C., Parisi, L. and Bottaro, D. (2003) Simple and rapid technique for monitoring the quality of meningococcal polysaccharides by high performance size-exclusion chromatography, *Journal of Biochemical and Biophysical Methods*, 56, 291–296.

Waldron, K.W. and Faulds, C.B. (2007) *Cell Wall Polysaccharides, Composition and Structure*, Oxford: Elsevier.

Willats, W.G.T., McCartney, L., Mackie, W. and Knox, J.P. (2001) Pectin, cell biology and prospects for functional analysis, *Plant Molecular Biology*, 47, 9–27.

Xie, G., Schepetkin, I.A., Siemsen, D.W., Kirpotina, L.N., Wiley, J.A. and Quinn, M.T. (2008) Fractionation and characterization of biologically-active polysaccharides from *Artemisia tripartita*, *Phytochemistry*, 69, 1359–1371.

Yamada, H. and Kiyohara, H. (1999) Complement-activating polysaccharides from medicinal herbs, in Wagner, H. (ed) *Immunomodulatory Agents from Plants*, pp. 161–202, Basel: Birkhauser.

Yamada, H. and Kiyohara, H. (2007) *Immunomodulating Activity of Plant Polysaccharide Structures*, Amsterdam, the Netherlands: Elsevier.

Yang, X.B., Zhao, Y., Wang, H.F. and Mei, Q.B. (2007) Macrophage activation by an acidic polysaccharide isolated from *Angelica sinensis* (Oliv.) Diels, *Journal of Biochemistry and Molecular Biology*, 40, 636–643.

Yapo, B.M. (2011) Pectic substances, from simple pectic polysaccharides to complex pectins—A new hypothetical model, *Carbohydrate Polymers*, 86, 373–385.

Yu, K.W., Kiyohara, H., Matsumoto, T., Yang, H.C. and Yamada, H. (2001) Structural characterization of intestinal immune system modulating new arabino-3,6-galactan from rhizomes of *Atractylodes lancea* DC, *Carbohydrate Polymers*, 46, 147–156.

Zou, Y.F., Barsett, H., Ho, G.T.T., Inngjerdingen, K.T., Diallo, D., Michaelsen, T.E. and Paulsen, B.S. (2015) Immunomodulating pectins from root bark, stem bark, and leaves of the Malian medicinal tree *Terminalia macroptera*, structure activity relations, *Carbohydrate Research*, 403, 167–173.

Zou, Y.F., Ho, G.T.T., Malterud, K.E., Le, N.H.T., Inngjerdingen, K.T., Barsett, H., Diallo, D., Michaelsen, T.E. and Paulsen, B.S. (2014a) Enzyme inhibition, antioxidant and immunomodulatory activities, and brine shrimp toxicity of extracts from the root bark, stem bark and leaves of *Terminalia macroptera*, *Journal of Ethnopharmacology*, 155, 1219–1226.

Zou, Y.F., Zhang, B.Z., Barsett, H., Inngjerdingen, K.T., Diallo, D., Michaelsen, T.E. and Paulsen, B.S. (2014b) Complement fixing polysaccharides from *Terminalia macroptera* root bark, stem bark and leaves, *Molecules*, 19, 7440–7458.

Zou, Y.F., Zhang, B.Z., Inngjerdingen, K.T., Barsett, H., Diallo, D., Michaelsen, T.E., El-Zoubair, E. and Paulsen, B.S. (2014c) Polysaccharides with immunomodulating properties from the bark of *Parkia biglobosa*, *Carbohydrate Polymers*, 101, 457–463.

Zou, Y.F., Zhang, B.Z., Inngjerdingen, K.T., Barsett, H., Diallo, D., Michaelsen, T.E. and Paulsen, B.S. (2014d) Complement activity of polysaccharides from three different plant parts of *Terminalia macroptera* extracted as healers do, *Journal of Ethnopharmacology*, 155, 672–678.

Storage Polysaccharides: Starch and Fructans

4

We now consider the structure and biotechnological use and potential of a group of polysaccharides responsible for storage of energy reserves: starch and its constituents (amylose and amylopectin), the related storage polysaccharide glycogen from animals and two furanose-based plant storage polysaccharides (inulin and levan).

4.1 STARCH

World production of isolated starch, in the form of grains and products made from them, is of the order of 20 million tonnes. The amount consumed as part of staple diets is very much greater than that and this is subject to attack by digestive enzymes in the gut. As will become evident in the following context, the extent to which it is attacked depends on the detailed structure of the starch and precisely, which enzymes and enzyme inhibitors are present. There is scope for biotechnology in improving starch, which is simply consumed as part of a foodstuff, as well as in the numerous processes which depend on a more or less purified starch as a raw material (Halley and Averous, 2014).

The most important sources of starch are maize, potatoes and wheat, with sorghum rice and barley as less significant localized sources. In some places, minor crops such as cassava are also processed to yield starch.

4.1.1 AMYLOSE AND AMYLOPECTIN

Students who studied biology at school will be familiar with these molecules. Amylose, amylopectin and also the animal storage polysaccharide glycogen (Section 4.2) are all homopolymers of glucose: that is, they are glucans. Amylose is a linear polymer involving $\alpha(1\rightarrow4)$ links whereas amylopectin (like glycogen) is a branched polymer containing an $\alpha(1\rightarrow4)$ linked backbone with regular $\alpha(1\rightarrow6)$ linked branches which can be up to 20 residues (each themselves linked $\alpha(1\rightarrow4)$) in length. Both polymers can be very large, particularly amylopectin (Harding et al., 2016): molecular weights for amylose have been reported between $0.2–2 \times 10^6$ Da and for amylopectin considerably higher ($\sim50–100 \times 10^6$ Da). Amylose is soluble in aqueous solvents where it exists principally in an approximately random-coil form (class D or C conformation type), with a Mark–Houwink a coefficient between 0.5 and 0.7 and a relatively small persistence length L_p between 2 and 4 nm. Under certain conditions, it can form a regular helical conformation, particularly when complexed with a stabilizing ligand such as iodine (Figure 4.1a). On the other hand, amylopectin (a class E conformation type – Figure 4.1b)

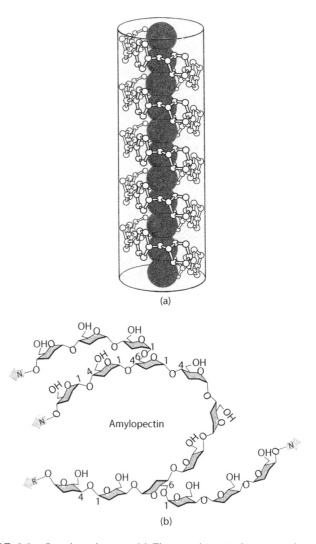

(a)

Amylopectin

(b)

FIGURE 4.1   Starch polymers. (a) The amylose–iodine complex: rows of iodine atoms (shown in grey) neatly fit into the core of the amylose helix. Unliganded amylose normally exists as an approximately random coil in solution (see text). (b) Part of the primary structure of amylopectin. The 'non-reducing' ends are indicated by an N and the reducing end by an R. (Reproduced from Mathews, C.K. and van Holde, K.E., *Biochemistry*, Benjamin/Cummings, Redwood City, 1990. Courtesy of Irving Geis©.)

is at least partially insoluble in aqueous media, but can be solubilized by solvents such as dimethyl sulphoxide or, like cellulose, ionic liquids.

4.1.2 STARCH BIOSYNTHESIS AND VARIANT FORMS

Although amylose is primarily a linear chain molecule, amylopectin (and to some extent amylose) is branched and specific enzymes are responsible for forming the branch point linkages. In every case, the glucose is activated by a reaction of the type

$$Glc + ATP = ADP\text{-}Glc + Pi$$

where ATP is adenosine triphosphate, ADP is adenosine diphosphate, Pi is inorganic phosphate and the ADP-Glc is added to a primer chain of $(Glc\text{-}Glc\text{-}Glc)_n$ with release of phosphate. Although ADP is the commonest activator, GDP and UDP are also used in what appears to be the common synthetic mechanism for all the cell wall and most of the storage polysaccharides. (There is evidence that higher plants contain two different starch synthase enzyme complexes, one using ADP-Glc and the other UDP-Glc (where UDP is uridine diphosphate), the latter possibly being an evolutionary relic. A similar situation seems to exist in lipid synthesizing enzymes.) The other common hexoses such as mannose, galactose, rhamnose and xylose and the pentose arabinose are all activated in this way, as are the uronic acids. With two exceptions, it is the universal mechanism. This is on the whole not good news for biotechnology. Reactions involving nucleotides have proved very hard to use in that context. Biotechnology tends to do well with relatively simple hydrolysis reactions, or at least has done so far in its development.

Fortunately, the exception may offer a way out of the difficulty. Fructans (see the following) are synthesized using a disaccharide as the activated donor, in this case sucrose donating a fructosyl residue. Thus,

$$Glc\text{-}Fru + Fru\text{-}Fru\text{-}Fru\text{-} = Glc + Fru\text{-}Fru\text{-}Fru\text{-}Fru\text{-}$$

and

$$Glc\text{-}Fru + Glc\text{-}Glc\text{-}Glc\text{-} = Fru + Glc\text{-}Glc\text{-}Glc\text{-}Glc\text{-}$$

avoid the use of nucleotide activation.

Another system apparently able to donate glucosyl residues from sucrose has also been reported. These look much more exploitable and attempts have been made to use them, and we will consider this when we discuss synthetic polysaccharides in Chapter 6. The other exception involves the synthesis of glycoproteins such as extensin, where a dolichol -P-P-polysaccharide chain is the donor when the chain is attached to the protein part of the molecule. The enzymes involved probably occur as a complex, within plastids in higher plants, and different enzymes are required for each sugar and for each specific reaction. Thus, it will require at least two enzymes to synthesize amylose from ADP-Glc, as well as the presence of a primer site.

Amylopectin differs from amylose in having many more branches, and the branching enzyme is relatively more involved. It has been clear for a long time now that there are a number of branching enzymes, which broadly accounts for the differences between different species in the extent of branching in their amylose and amylopectin. This is not unreasonable. If one considers a chain in the process of extension, at some point a branching enzyme intervenes. This is likely to be determined by the length of the existing chain which probably reaches a point at which the ability of the branching enzyme to bind to it exceeds that of the extending enzyme. However, as soon as a branch point is inserted the substrate becomes a quite different kind of molecule with a sharply different shape. A second branching enzyme could well be better adapted to this shape. The same argument can be pursued to a second branch point. At least three branching enzymes are known to exist in maize, which has an interesting series of mutants that have helped to elucidate the pathways. A summary of some known mutants affecting starch is given in Table 4.1.

4.1.2.1 STARCH APPLICATIONS

Starch has a number of applications, and a substantial proportion of it is used in non-food ways, and these applications can depend crititically on the relative levels of amylose and amylopectin. Paper coating is one example: at present, potato starch has to have its viscosity modified by hydrolysis

TABLE 4.1   Mutations Involving Starch Synthesis

Species	Change	Enzyme Involved
Maize		
Waxy	No amylose, only amylopectin	Starch 'synthase'
High amylose	70% amylose	Branching enzyme 2b absent
Dull	55% amylose	Branching enzyme 2a absent
Sugary	Highly branched amylose (phytoglycogen)	Branching activity enhanced
Shrunk	30% normal starch	Not known
Barley		
Waxy	Amylopectin only	Starch 'synthetic' enzymes
High amylose	43% amylose	Not known
Notch 1	50% reduced starch content	Not known
1508 Bomi	Small starch grains	Not known
Rice		
Waxy	1.5% amylose	Not known
Pea		
Waxy	Amylopectin only	Starch 'synthase'
High amylose	70% amylose	Branching enzymes
Arabidopsis thaliana		
Ethylmethylsulphonate mutants	No leaf starch	ADP-Glc synthesizing system

before it can be used and it is thought that a high-amylopectin starch could be used more directly with less preliminary treatment. High-amylopectin starches are less sensitive to the presence of ions such as calcium and sodium, which can adversely affect the viscosity of ordinary starch.

4.1.2.2 STARCH GRAIN SIZE DISTRIBUTION

Starch granules derived from waxy maize and a high-amylopectin maize, *amylomaize VII*, have been investigated for the sizes of the fragments of amylose and amylopectin produced when the grains are exploded by microwave heating. It is possible in these varieties to image the particles by electron microscopy and a representation of the results is given in Figure 4.2. As

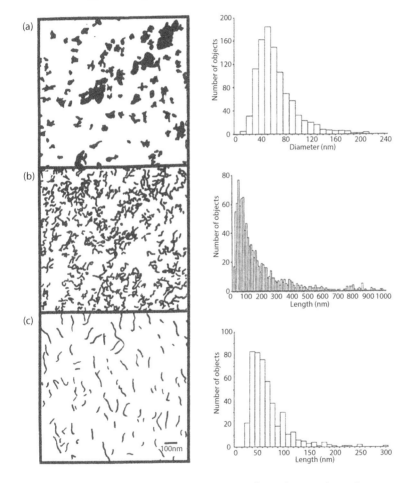

FIGURE 4.2 Electron microscope images of amylose and amylopectin obtained with samples made by microwave heating of a suspension of maize starch in water. (a) Amylopectin, (b) linear amylopectin and (c) amylose pseudohelices. To the right is shown the corresponding size distributions of the particles. The amylose probably consists of an aggregate of six molecules on average. (Reprinted from *Carbohydrate Polymers*, 26, Fishman, M.L. et al., Size distributions of amylose and amylopectin solubilized from corn starch granules, 245–253, 1995, with permission from Elsevier.)

expected, amylose gives linear particles with some appearance of helicity, while amylopectin showed as either space filling patches or linear particles. About 30% of the fragments were in the circular form with sizes up to 200 nm. The average particle appeared to contain about six molecules and had an effective molecular weight of about 2×10^6 Da. These results are comparable with those from gel-filtration methods, and are relevant to the kind of particles important in food formulations.

The formulation of quick-cook packet soups and similar products depends on having a very consistent range in the raw materials and the wrong size of grain can ruin the product. Mutants are known where the grains do have a different size distribution, but the reasons for this must lie in complex interactions between the quantity and spatial distribution of the synthases as well as supply of substrates. It is rather remarkable that a single mutation can apparently have such an effect, and indicates the huge scope for manipulation potentially available in starch synthesis. If the same rules apply to heteropolymers, and they probably do, there are clearly possibilities of making novel materials. As well as all this, the presence of starches at high levels in the seeds of many species means that cheap and easy production should be possible.

Demand for such materials remains much more problematic, and until and unless some uses are identified there does not seem to be much scope for production of modified starches by direct manipulation of the biosynthetic pathways. Production of a range of materials from starch by biotechnology processes is a relatively mature industry and most of the niches have been filled by currently available materials. However, novel requirements may well be identified, for example in the cyclic dextran and related compounds and their applications, which if they are ever made on a large scale will almost certainly be made from starch.

One raw material, xanthan gum (Chapter 6), was originally conceived in 1959 as a way of using starch to produce more interesting materials, in this instance by using it as a medium for *Xanthomonas* growth in fermenters. Specialist baking may also lead to new demands.

4.1.3 STARCH DEGRADATION

The enzymes that degrade starch are some of the most commercially important in current biotechnology both as crude isolates and *in vivo* in the various brewing processes based on cereals. These enzymes account for about 25% of sales of enzymes by value, exceeded only by total proteinases, most of which would be due to enzyme detergents. For food use they are probably the most important in all the significant markets (i.e. Europe, Japan and the United States). In tonnages too, they are prominent but tables of enzyme production are of little comparative value because the actual enzyme content of the dry powders usually sold is variable and low. There is not much point in noting that 300 tonnes of *Bacillus* amylase was sold without also knowing what the actual enzyme content was. This is often very low, usually below 1% of the protein, but it is possible that the 10 tonnes of fungal amylase also sold actually contained more enzyme. Using units of enzyme activity does not bring much improvement since these are quoted in units that are often arbitrary and not comparable between different preparations and different applications.

The objective of enzyme digestion of isolated starch varies. Partial hydrolysates have found many applications and still constitute the most important products. One enzyme process leads to maltose, while a whole variety of dextrans are used in food applications. The ultimate product of hydrolysis is glucose, and another target is to convert as much as possible into it – and of more recent interest – followed by the partial conversion of the glucose to fructose.

The traditional process was acid hydrolysis, followed by neutralization. This had drawbacks – neutralization invariably causes the appearance of a salt, usually sodium chloride, which may be unwanted in later applications. Also acid hydrolysis is not complete except under fairly extreme conditions of pH and temperature, which leads to the formation of undesirable side products. Starch grains, which would be the usual raw material, do contain small amounts of protein and lipid which become involved.

Thus the use of enzymes, under mild pH conditions, though surprisingly enough routinely at 100°C, avoids most of these problems, and an enzyme process for partial hydrolysis has been used for more than 50 years. It should be emphasized that complete conversion of starch to glucose is difficult and requires a number of enzymes acting in concert. This is not surprising given the extensive branching of the structure. A combination of enzymes is now available which can yield a wide range of hydrolysis products, though it is probably premature to say that any desired mixture could be produced.

If any target should become well defined and sufficiently valuable it might justify attempts to modify the synthesis of starch to confine it to unbranched amylose chains only. While this might help it would not, as we shall see, be the complete answer. The problems of obtaining complete hydrolysis are really inherent in the fact that it is a long polymer that can be attacked either randomly or at either end, which will inevitably lead to an asymptotic approach to completion. Reference should be made to the discussion of cellulose and cell wall breakdown we have already given in Chapter 2 and the role of multiple and fused enzymes.

The same problem is found with proteolysis where complete enzymatic digestion of proteins is almost impossible. This is partly due to the differing specificities of the proteases and the wide variety of amino acid–peptide links, but even when a suitable collection of proteases is used to complete hydrolysis is not achievable. If it *were* possible, it would be very useful for exactly the same reasons as for polysaccharides. An alternative approach might be to find enzymes that break all the bonds in starch efficiently, so as to bring about at least a near approach to complete hydrolysis. This is broadly the approach that has been followed. Where the starch is to be broken down *in vivo*, as in brewing, combinations of enzymes can also be used. In modern conditions it is to be expected that suitable enzymes, in whatever organism they might be found, could be transferred to a suitable production organism and made available. In brewing (such as *sake* manufacture, where two organisms are used because they each produce a starch degrading enzyme that together result in effective degradation), it has proved possible to gather the enzymes together, by gene transfer, into one organism which can do the *whole job*. One could also envisage this being done where a mixture of starch degrading enzymes is needed so that they could all be obtained from a single source organism. In practice, most products made by hydrolysis of starch are well short of complete breakdown to glucose, and have already found wide application in the food industry.

4.1.3.1 STARCH GRAINS AND HETEROGENEOUS ATTACK

In most organisms, and all those used as commercial sources, starch occurs in the form of insoluble grains. These vary in size and shape but usually have radial symmetry, which is the reason for their spectacular appearance in polarized light, and a highly ordered structure. On germination starch grains are used up and disappear along with the other storage materials of the seed. They may also appear and disappear during the maturation of the seed. The soybean for example has many starch grains in the cotyledon until just before maturation when all the starch is converted to lipid. The mature seed contains none. This can happen quickly and there are clearly very active enzyme systems able to break down starch grains.

Classical biochemistry was almost entirely concerned with enzymes acting on substrates in solution, and not much is known about the way in which enzymes attack lumps like starch grains. A similar situation in understanding the mode of action of lipases has only recently been improved, under the stimulation of a sharp increase in the uses of lipases in biotechnology (Jaeger and Eggert, 2002).

In the case of amylases, one innovative study has used the technique of *fluorescence recovery after photobleaching* ('FRAP') to follow the movement of amylase molecules on a starch matrix: Henis et al. (1988) found that the technique could be used to follow the lateral diffusion of labelled β-amylase molecules over the surface of a starch gel. There were two components to the motion. One depended on an exchange between bound and free enzyme molecules while the other was a simple lateral motion. The lateral diffusion disappeared when the enzyme was inactivated by reaction with iodoacetamide, suggesting that activity was linked in some way to motion, though not to simple binding.

When starch grains are heated they swell and gelatinize. The temperature of gelatinization varies with the particular grain used, from 50°C for potato up to 68°C for sorghum, rice and high-amylose maize. There is a correlation with the amylose content. The result is an amylose gel containing fragments of unexpanded parts of the original structure in an irregular way. Eventually, a viscous mass called a starch paste is produced. The transition from an ordered structure to a random gelatinous one actually creates opportunities for interactions between the molecules which can then form either an insoluble white precipitate or in more dilute systems an elastic gel. This phenomenon is called *retrogradation*, and probably makes access of degradative enzymes more difficult. In a different context it is part of the staling mechanism of bread. The hydrolytic enzymes are added to this mass and fortunately since they can operate at temperatures up to 110°C, it is not necessary to cool the slurry first. The substrate is physically heterogeneous and quite apart from the problems of enzymes in random attack on long polymers, access to retrograded particles must be different from those on gelatinous ones.

Thus, there appear to be at least three starch degrading enzyme systems. The first of these is that which is involved in the balance between lipid and starch content of the seed. Seeds contain both lipid and starch as storage materials, though the relative amounts vary greatly. Some such as groundnut contain both, while others such as peas may contain starch but little lipid, and mature soy has no starch but contains it earlier in the maturation process. The switch controlling this level is near to CoA carboxylase in the metabolic pathways, and in effect determines the fate of the acetyl CoA pool, but there is clearly a mechanism by which starch is broken down to acetyl CoA *in vivo*. It is likely that this is a simple reversal of the synthetic pathways.

Second, there are the enzymes responsible for breakdown on germination and mobilization of the reserves. Dry mature seeds contain little enzyme activity and there is accumulating evidence that an early event in germination is the appearance of mRNA coding for the hydrolytic enzymes. These are clearly able to bring about hydrolysis, and are exploited in the malting process in brewing, and to a lesser extent in baking.

Third, there are the enzymes associated with digestion. Saprophytic organisms such as fungi and bacteria all have them. They are effectively digestive enzymes on a par with the amylases of the mammalian gut. There are many thousands of species producing such enzymes which probably all belong to a small number of families and exist in hundreds of variants. It is doubtful if they are specialized to produce complete hydrolysis. Because fungal and bacterial culture media are convenient sources for the production of enzymes, most of those which have been used so far in biotechnology applications are of this type. The extent of degradation of starch in the human gut is of current interest because it is thought to be a factor influencing the fibre content of the diet. Although outside the scope of this book, this is one influence which might lead to demands for a more controlled level of branching in dietary starch. There is also some interest in non-ruminant animal feeds in the so-called resistant starch as a component of the diet.

The enzymes used *in vivo* for germination of the seeds have not been exploited (other than indirectly in the malting process) and, since they presumably attack the highly ordered starch grain in the cell, the question of whether they could attack an isolated grain without preliminary gelatinization might be asked.

Some studies on degradation *in vivo* have shown that the attack is heterogeneous and leads to holes in the starch grains, and can in some instances even lead to the development of a hollow shell. Corn starch granules show similar effects when treated with bacterial α-amylase *in vitro*. It is believed that the potato starch granule is the most resistant, but even this is extensively digested by a *Bacillus circulans* amylase which yielded maltohexose – though an amylase from *Chalara paradoxa* is claimed to yield only glucose. Many starch grains have small pits on the surface which contain protein, and there is some evidence that these are connected with the initial attack and hole formation. In some instances, the attack seems to be by general erosion.

A range of new products based on partial hydrolysis of starch grains that have not been gelatinized, or only partly gelatinized, have during the last few decades become increasingly popular. Called 'starch hydrolysis products' to distinguish them from maltodextrins and corn syrups, they have properties that appear to depend on the possession of higher molecular weights (Woods and Swinton, 1995).

During storage of potato tubers, some of the starch is converted to sucrose and reducing sugars when storage is at less than 5°C. This is reversible and starch is reformed when the temperature is raised, and although the question of whether this represents the reversible synthesis of a few starch grains, or the partial degradation of most of them, has been raised, it has not yet been answered. Both phosphorylase, which would be involved in any reversal of synthesis, and α-amylase are present at significant levels in the mature tuber, though little is known about the debranching enzymes that would be essential to hydrolytic breakdown. The amylase level does not increase much on budding, but little is known about starch degradation at this stage. In contrast, the levels of amylase activity in fruits and seeds do increase substantially on germination, and potato tubers may be an exceptional situation. Prolonged storage of tubers at ambient temperature also leads to a slow accumulation of reducing sugars, probably by amylase action.

Starch breakdown in cereals is under hormonal control, and the giberellins are responsible for the production of a series of amylases that are able to break up the starch granules. However, phosphorylase is also prominent and it is also involved. The overall pattern which emerges is one where hydrolases such as α- and β-amylase break down the granules, and partially break up the amylose and amylopectin into fragments which are then available for phosphorolysis. Thus, the true function of amylases in the storage organs may be to break down the insoluble starch grains so as to render them accessible to the enzymes such as phosphorylase which would channel them into the metabolic pathways of the growing organism.

It is possible that this is also the function of the 'digestive' amylases, which are intended to make the otherwise intractable starch break down to absorbable fragments, rather than to take them all the way to mono- or disaccharides. The ability of amylases to attack intact starch grains, of their own or other species, with apparently variable results is a factor which should always be taken into account when choosing a suitable enzyme. It is likely that different amylases will give widely different results, and a variety should be evaluated.

4.1.4 ENZYMES OF STARCH DEGRADATION

Some individual enzymes which are important in starch processing are as follows.

4.1.4.1 α-AMYLASE

This enzyme (or group of enzymes – see Zeeman et al., 2010), which is very widely distributed and is an important digestive enzyme of the human gut, breaks the (1→4) linkages in both amylose and amylopectin. It does this randomly, acting mostly as an endohydrolase, and cannot break the (1→6) branch point links. Also, presumably because of steric hindrance effects, it stops short of the branch points by one or two residues to yield the so-called limit dextrans. The α-amylases of *Bacillus subtilis* and *Bacillus licheniformis* are commonly used, have the useful property of high stability at 100°C and are used at this temperature in processing. This has the advantage of maintaining sterility, as well as a faster reaction. It is known that proteins are stabilized against thermal unfolding in the presence of polyols such as hexoses (Tombs, 1985) and it is possible that this is a factor in the unusual stability of these enzymes. If so other enzymes should also show greater stability in these highly concentrated systems.

A typical out-turn of extensive α-amylase activity would be about 5% glucose, 50% maltose and 30% maltotriose, with significant amounts of higher polymers also present as well as the limit dextrans whose amount will vary with degree of branching of the particular starch chosen. They are typically around 20% of the total. Clearly using this enzyme alone is not the way to produce glucose. The presence of α-amylases has been reported in numerous organisms and the choice of the commercial producing organism is largely a matter of convenience. There will be many different sequences found amongst the enzymes though it does not follow that every species has a unique sequence, and the number of different enzymes is likely to be smaller than the total number of species producing them. From the viewpoint of the biotechnologist, the main factor determining the choice of amylase is likely to be the ease of production and general availability rather than any specific property of an individual amylase. There may be local specialization. For example, recently in Nigeria an amylase from *Aspergillus niger* has been described that is good at breaking down intact starch grains from sorghum, cassava and maize. Isolated from a cassava rotting strain it may well find local uses where it can outperform amylases adapted to rice or barley.

4.1.4.2 β-AMYLASE

This enzyme, which attacks only the α(1→4) linkages and like α-amylase cannot attack the branch links, is very widespread and again like α-amylase has the status of a digestive enzyme. It attacks only the ends of the chains to release maltose. That is, it attacks only the penultimate link. It occurs together with α-amylase and forms part of the digestive battery of enzymes. The two together produce almost entirely maltose and the limit dextrans. This combination is strongly reminiscent of the synergy between carboxypeptidases and endopeptidases in digestion and is a neat solution to the problem of breaking down long polymers. The product, however, is maltose, and the combination is still not the answer to glucose production.

4.1.4.3 GLUCOAMYLASE

This enzyme, like amylases attacks the (1→4) linkages in the exomode, releasing glucose, but its main interest is that it can also break the branch point (1→6) linkages. Thus, it increases the yield of glucose and is used after an initial treatment with α-amylase. They are not used together since glucoamylase cannot withstand temperatures above 60°C and the pH should be around 4.5 for optimum activity. A combination of amylase and glucoamylase activity can result in products with 95%–98% glucose which are widely used in food applications, as well as forming the feedstocks for high-fructose syrup production (see the following).

4.1.4.4 ISOAMYLASE

This (1→6)-linked α-D-glucan maltohydrolase removes maltose from the non-reducing ends of amylose and amylopectin. It is found in bacteria, particularly *Pseudomonas* spp. It is effective in combination with β-amylase in raising the level of maltose.

4.1.4.5 PULLULANASE

An endo-α(1→6)-linked glucosidase, this enzyme is important because it splits the (1→6) links of amylopectin. It is obtained from *Bacillus* species and derives its name from its ability to hydrolyze pullulan (Chapter 6), a linear polysaccharide made up of maltotriose units linked by α(1→6) linkages found in *Aureobasidium pullulans*. There are several hydrolytic enzymes found in this organism, and some thermostable pullulanases occur in *Clostridium thermohydrosulphuricum*. These can break α(1→4) bonds in starch and could be used in the liquefaction step since they can withstand 90°C.

4.1.4.6 PROTEIN ENGINEERING OF THE AMYLASE FAMILY

All the enzymes mentioned earlier belong to the same family, as judged from their sequences and several crystallographic structures are currently available. Table 4.2 lists some of the known structures. Study of these structures leads to the conclusion that all members of the family have a barrel-like arrangement of eight β sheets and eight α helices, shown as $(\beta/\alpha)_8$. The α-amylases have a domain between the third β strand and the third α helix, as shown in Figure 4.3. The catalytic and substrate binding residues are at the C ends of the β strands, and in loops in the same regions binding sites have been identified in other locations as well. In particular, the starch binding domain is found near the C terminus.

Numerous sequences are now available representing 18 different specificities in starch hydrolysis, and show conservation in four or five regions. These include the catalytic glutamate, arginine and histidine,

TABLE 4.2   Some Starch Hydrolases of Known Three-Dimensional Structure

Enzyme	Organism
Taka-amylase	*Aspergillus oryzae*
Pancreatic α-amylase	Pig
α-amylase	*Aspergillus niger*
α-amylase type 2	*Hordeum vulgare* (barley)
Cyclodextrin glucotransferase	*Bacillus circulans*
Cyclodextrin glucotransferase	*Bacillus stearothermophilus*
(1→6) glucosidase	*Bacillus cereus*
β-amylase	*Glycine max* (soybean)
Glucoamylase	*Aspergillus awamori*

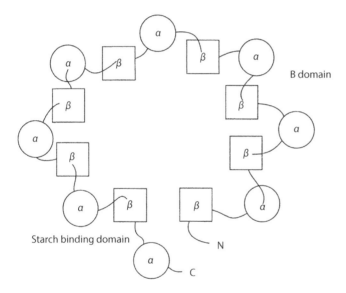

FIGURE 4.3   Schematic structure of the α-amylases and related enzymes. They are based on the α_8/β_8 arrangement, i.e. have eight α helices and eight β sheets alternating around a drum shape and have a starch binding domain near the C terminus and another characteristic domain between the third α and β structural elements. (Adapted from Svensson, B., *Plant Molecular Biology*, 25, 141–157, 1994. With permission.)

an important structural glycine, and a variety of substrate binding sites. Thus, all the detail needed for protein engineering is available, and has been employed in studies on α-amylase, cyclodextrin glucotransferases and pullulanase.

The α-amylase of *Bacillus stearothermophilus* and a number of other organisms has had its detailed specificity altered by directed alteration of residues in the active centre and the binding site. In most cases, as might be expected, alteration of active centre residues led to loss of activity while alteration of binding sites led to changes in reaction rates. In one case, *Saccharomycopsis fibuligera* mutants showed a small change in preference for longer as against shorter chain substrates. The human pancreatic enzyme has enhanced maltase activity when the His210 residue is altered to asparagine. Other attempts to alter specificity from amylase to pullulanase type on a rational basis from sequence comparisons were rewarded with substantial loss of activity.

The amylase of barley has been much studied because of its commercial importance. Cereal amylases bind β-cyclodextrin to a site that is probably the same as the starch grain binding site, since it competes with it. It has a pair of tryptophan residues Tyr278-Tyr279, and changing 279 to alanine results in much lower cyclodextrin binding and an enhanced binding to starch. Residue 278 could not be altered with retention of structural integrity and is invariant in cereals. When barley amylase is expressed in yeast, one of the four cysteines is converted to a mixed disulphide with glutathione (so is about one-third of human albumin in man). This results in inactivation, apparently because Cys95-glutathione prevents proper folding. There is no reason to expect this to be a general effect of glutathione attachment, though it is interesting that yeast can do it in heterologous expression. There are two barley amylase isozymes, with only about 80% sequence conservation, with distinctly different specificities. Hybrids have been made and expressed in yeast but the results are complicated and not fully interpreted as yet. A number of different hybrids can be made with varying results.

There are so many possibilities that without some better idea of the targets for site-directed mutagenesis and the use of mutants in general it is difficult to make use of the mass of detail that is beginning to accumulate. It is clearly going to be possible to obtain an array of enzymes with well-understood and defined specificities, but just how they might be used remains obscure.

4.1.4.7 GLUCOSE ISOMERASE

This enzyme is used in the final stage in the conversion of starch to fructose, and while it is not strictly speaking involved with polysaccharides it cannot be ignored. When the idea of converting glucose to fructose, as an alternative sweetener to compete with sucrose, first arose the best-known glucose isomerase was the mammalian enzyme. This involves a phosphorylation reaction and ATP. As already pointed out, reactions involving nucleotide cofactors are just not feasible, and it was not until some bacterial enzymes capable of carrying out the isomerization were found that an enzyme-based process could be developed (see the following). The enzyme in question is probably a xylose isomerase, and a number of sources have been found and approved for food use. It is relatively expensive to produce which makes it worthwhile developing a column-based immobilized enzyme system, so that it can be re-used as much as

possible. In practice, the enzyme is used in the form of the disrupted producing organism. The enzyme appears to adhere to the cell fragments in
an active form.

4.1.5 PROCESS SCALE STARCH DEGRADATION

4.1.5.1 STARCH GRAIN ISOLATION

The first step is the preparation of clean starch grains from the source
material. The process will naturally vary with the source. In a typical 'wet milling' operation, based on maize, the grain is first soaked
in water for 48 hours at 50°C. This softens the grain, and also extracts
soluble carbohydrates, lactic acid and minerals to yield corn steep
liquor. This is an important growth medium for fermentations and is
widely used for antibiotic manufacture. Processing of grains and beans
invariably leads to the production of large volumes of more or less
dilute solutions of carbohydrates, often pentosans which are very difficult and expensive to dispose of. Plants processing many thousands
of tonnes of grain a year require, nowadays, dedicated effluent treatment plants to handle what remains a growing problem. The economics of waste disposal are now a very significant part of the overall costs
of processing plants. Anything which can reduce the cost of disposal,
preferably by finding a use for the waste materials, is highly desirable. A
substantial cattle food industry has grown up based on the by-products
of oilseed processing, where the 'meal' remaining after lipid extraction
has found value as an animal feed. However, much remains to be done
since some of the waste products cannot be used, even at zero value,
as cattle food.

The next step is mechanical removal of the hypocotyls, and lipid
extraction by pressing or solvent extraction. The residues are then
ground, and are effectively a mixture of cellulose cell wall fragments,
protein bodies – the so-called 'gluten' – and starch grains. They are
separated by a flotation and centrifugation process which takes advantage of the high density of starch grains, to yield a suspension which
is almost entirely starch. It contains, as does the starch grain, about
1% protein and a similar amount of lipid. It is drum dried for storage
and transport, but increasingly is further processed in the same factory
without drying.

The details of the processing are very specific to a particular crop. It
is said for example that wet milling is not suitable for European grown
varieties of maize. There are possibilities of using enzymes during wet
milling, though this mostly involves cellulases as an aid to cell wall
disruption. There are also possibilities of modifying the cell wall by
suitable mutations with a similar objective. Several hundred thou-
sand tonnes of maize are processed in this way, but even this is only a
small proportion of the total maize production and it may be difficult
to justify growing a special variety of maize just for this application.
Since the process is already highly efficient, the gains would need to be
substantial.

4.1.5.2 ENZYME DIGESTION

Starch is mixed with water to give a 30%–35% suspension (at high-starch contents the mixture is thixotropic, and this has been exploited as the supporting medium for a preparative-scale electrophoresis method). There is little possibility at this stage of using column-based continuous methods because the mixture is too viscous and all the processes are based on batch methods with no attempt made to recover the enzymes. The tank containing usually at least 500 L is then steam heated, the pH adjusted to between 6 and 7 by addition of lime, and an amylase preparation added. The most widely used one is from *B. licheniformis* and has the unusual property of long-term stability at 100°C despite the fact that the source organism is not a thermophile. It is activated by calcium ions, and if the pH is not adjusted with calcium hydroxide, then these must be added. Steam heating causes gelatinization of the starch but the simultaneous action of the enzyme causes a thinning out which reaches completion in about 3 hours. The amount of enzyme added, expressed in terms of some convenient activity measure, is adjusted empirically so as to produce this result with minimal enzyme content. Loss of activity is believed to be mainly due to hydrolysis of amide groups from the enzyme asparagine and glutamine residues. The now popular starch hydrolysis products (SHPs) are made in the same way but without such extensive gelatinization and are less extensively hydrolyzed.

Some processes stop at this point and the *maltodextrins* are sold as such, and are characterized by their 'DE' value (this means 'dextrose equivalent' and *not* the 'degree of esterification' used in pectin terminology): the reducing power as a proportion of that which would be shown by pure glucose. It is a rough estimate of the number of reducing ends released by the enzyme attack. The higher the number, the lower the average molecular weight. Obviously, the precise mixture of dextrans will depend on the nature of the starch used, particularly the degree of branching, but it is said that the precise α-amylase used can also affect the outcome.

An alternative treatment makes use of fungal α-amylases, often from *Aspergillus oryzae*, which release maltose and produce syrups with a relatively high maltose and lower oligomer content; β-amylases and pullulanases may also be used on the starch slurry, but in all cases the temperature must be reduced to 50°C since they do not have the temperature resistance of bacterial α-amylases. The solution is clarified by filtration and usually dried to a solid content of 65%–75% and stored in drums. Thus, a variety of mixtures of maltose and higher oligomers is available and used mostly on a semi-empirical basis. Nearly complete conversion to maltose can be achieved with these enzymes. Some typical compositions produced by the various methods of hydrolysis are shown in Table 4.3 and an outline of the way in which processes have developed since 1950 is shown in Figure 4.4.

More recently, there has been a demand for more extensively hydrolyzed starch as a substrate for conversion to fructose.

TABLE 4.3   Starch Hydrolysis Methods

	Products of Starch Hydrolysis (%)			
Method	Glucose	Maltose	Triose	Oligos
Acid	91.5	2.5	6.0	
Acid enzyme	93.0	2.5	3.5	
Enzyme	96.0	1–2		0.1
Maltose/ dextrose	18.0	43.0		
High maltose	5.0	52.0	15.0	

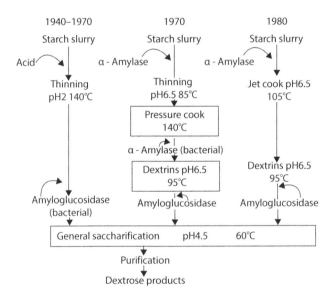

FIGURE 4.4   The development of starch degradation processes since 1940. The objective is to produce glucose or a range of polyglucose syrups. (Adapted from Cowan, *Biotechnology: The Science and the Business*, Gordon & Breach Publishers, Lausanne, Switzerland, 1991. With permission.)

4.1.6 GLUCOSE AND FRUCTOSE MANUFACTURE

The liquefied starch is cooled to 50°C–60°C and the pH adjusted to 4–4.5 with hydrochloric acid. Then a suitable glucoamylase is added and the mixture maintained for 2–3 days, depending on the amount of enzyme added. Liquefied starch which has been treated with β-amylase or pullulanase can also be used as a feed for this stage. The *Aspergillus* enzyme is most often used and leads to a syrup which is 96% glucose. It is usual to refine the syrup by charcoal adsorption to remove small amounts of colouring material, and it is then sold as a concentrate or crystalline material. The major use for glucose syrup is now for conversion to high-fructose corn syrup (HFCS). In 1985, about half of the world production was converted to fructose: most of the remainder went into confectionery.

The rise in demand for fructose syrups is a reflection of the worldwide changes in sweetener consumption. Sucrose which in terms of tonnages is pre-eminent has for some time been partly displaced by fructose for a variety of reasons (Tombs, 1990). Fructose is sweeter than glucose, has a higher solubility which makes it more convenient to use and is able to compete with sucrose both in terms of price and applications. This has been particularly so in the United States where at least a billion dollars worth of imported sucrose has been displaced. In other parts of the world, fructose has been held back by deliberate support for sucrose for political reasons, and it is still unclear whether fructose can compete in a completely free market. The increase in obesity in many populations has been blamed on an increase in fructose consumption although some studies have suggested this not to be the case (Akhavan and Anderson, 2007). Since it is based on what is frequently a by-product, that is starch, from some other process, it might be able to. At present, the starch raw material makes up about 50% of the total costs of making HFCS. It should also be noted that there are other potential sources of glucose – for example from cellulose hydrolysis – and fructans (see the following) are a source of fructose and in principle glucose as well, since the glucose–fructose conversion is reversible. Milk whey is a notorious waste product and attempts have been going on for many years to find a use for it. The lactose can easily be converted to a mixture of glucose and galactose by an immobilized lactase, but so far it has failed to make any impact on starch produced glucose. Instead, it has been fermented to ethanol but only in pilot-scale operations.

There are a number of glucose isomerases, most of which require nucleotide cofactors and could not be used in a large-scale process. Eventually, a xylose isomerase was found that could also isomerize glucose to fructose and did not require nucleotides. The enzyme is difficult to produce, which has led to the use of continuous flow columns for this part of the starch-to-fructose process. Much work has been done to optimize the enzyme source, and it is now made by *Streptomyces olivaceous*, though other bacteria including *Actinoplanes missouriensis* have also been approved, on a glucose source. Xylans can be used to induce the enzyme. It catalyzes the appearance of the equilibrium mixture of 55% fructose 45% glucose. Higher fructose levels have to make use of a chemical fractionation of fructose from glucose; fructose syrups with more than 55% fructose slowly revert to the equilibrium mixture.

Since HFCS is used in foods it is important that food standards are maintained, and the enzyme sources must be approved.

The packed columns have a half-life of 70–100 days, and the substrate is run in at about 50% solids with a residence time of just a few minutes. This is enough to give a yield of 42% fructose, a suitable compromise between yield and conversion time. An experimental arrangement has been described in which starch is kept on one side of an ultrafiltration membrane and treated with a suitable mixture of hydrolytic enzymes. The small products such as glucose can then pass through the membrane and could be produced continuously. For the time being, however, the conversion of starch to fructose remains a mixture of batch and continuous column processes.

4.1.7 BAKING TECHNOLOGY

Baking is a process in which flour, usually obtained from wheat, is mixed with water and then fermented with yeast, followed by heating to make bread. Many variants exist, some of them based on other cereals, and some of them not using yeast. In all cases, however, the endogenous enzymes of the cereal are involved, and in our case the α-amylase and the β-glucanase are most important. The properties of the dough produced by mixing flour with water tend to be dominated by the protein components, the gliadin and glutenin. The status of the starch grains became important in a quite unforeseen way when the United Kingdom was negotiating to enter the European Economic Community. It was anticipated that this would lead to a decreased use of hard North American wheat, and an increase in the use of soft continental wheats. The former were believed to lead to a longer shelf life, while the latter notoriously produce bread that stales very quickly. Thus, the nation's daily bread was at risk, and much was made of this by politicians. North American wheat is characterized by a glutenin–gliadin ratio different from continental varieties, and a number of research projects were started to see if adjustment of this factor could be used to overcome the perceived difficulties. In fact, it turned out that the harder wheat led to different grinding requirements, which in their turn led to greater fragmentation of the starch grains, and this is the factor that determines the staling rate. It is common practice to modify the state of the starch grains by adding additional amylases. Malt was widely used, but since it also contains many other enzymes greater control is desirable. This has been obtained by using bacterial and fungal α-amylase from *A. niger* and *B. subtilis*, both of which are permitted food additives. They produce an improved colour as well as improved crumb structure. Addition of amylases does retard staling, but the structure is due to better gas production by the yeast during the fermentation stage. The key function of the enzyme is the way in which it attacks intact starch grains and modifies their properties, as well as the level of fermentable maltose it produces.

Glucoamylases can also be used to boost yeast growth and loaf volume, though this enzyme is not yet widely used. It should be noted that addition of xanthans and guar gums also retards staling, and there are undoubtedly physical effects to do with the movement of water involved in what is a complicated process. There are substantial unsolved problems in microwaving bread-based products, which are also dependant on the status of the starch grains.

Pentosanases (or hemicellulases) are also widely used in baking practice. They derive from *A. niger* and *Trichoderma viride*, and are present as minor activities in standard α-amylase preparations (indeed it is thought that some of the effects of amylases are really due to the contaminating pentosanases). Commercial materials are anything but homogeneous, and contain a number of different though related activities. Wheat contains fairly high levels of pentosans, which are insoluble in water, but bind it. Hydrolyzing them leads to a reduction in bound water and increase in the free water, which in turn leads to softer doughs. While there is a clear improvement in loaf volume and general properties, it is not possible to go much further than that as an explanation of the effect, which remains obscure. There is some evidence for complex formation between glutens and pentosans, which would certainly affect the dough structure. Pentosanases are especially important in controlling the water binding in high-fibre breads, where water binding is more difficult to control.

4.2 GLYCOGEN

Glycogen is the characteristic storage carbohydrate of animals, and in higher organisms is found mainly in the liver and musculature. Glycogen-like materials are, however, also found in plants and it seems to be unusual compared to other polysaccharides in having a compact spheroidal structure (Morris et al., 2008). It is an $\alpha(1\rightarrow4)$ glycan, as is amylopectin, but differs in being much more highly branched. Like starch it is synthesized and degraded *in vivo* by way of a phosphorylase, to give glucose-1-phosphate. There is little or no biotechnological interest in glycogen as such – and it is highly unlikely that animal sources would replace plant in the foreseeable future.

A major interest is in a series of rare human hereditary variants, in which branching enzyme deficiencies lead to the synthesis of glycogens with different structures. They have close parallels to those found in maize mutants and were useful in disentangling the various enzyme activities. The best-known condition is *Andersen's* disease, resulting from mutation of the GBE-1 gene which leads to linear unbranched glycogen being formed and in *Pompe's* disease where an $\alpha(1\rightarrow4)$ glucosidase activity deficiency leads to deposition of glycogen in the lysosomes. In the latter case, attempts to alleviate the condition by administration of the enzyme in liposomes have had only limited success. While this may be a candidate for gene replacement therapy it is not likely to be available in the near future.

4.3 MINOR STORAGE POLYSACCHARIDES

Starch is the most important storage polysaccharide, and has the all-important property of insolubility in water which is a characteristic of all storage materials. Nevertheless, many other polysaccharides are found at lower levels in the storage organs, and some of them are water soluble. They should probably be considered as storage materials too, if only because they have no other obvious function. Some of them are mainly located in the cell wall and might contribute something to its integrity, but again the general view is that they also have a storage function.

They are mostly neutral sugars, and are polymers of galactose, arabinose, rhamnose, mannose and glucose, together with small amounts of other sugars. They are very widespread. These, like glycogen, have no special interest for biotechnology and are best known rather for the problems that they can cause. This is because they tend to turn up in the effluents from seed processing and are mainly responsible for the extremely high biological oxygen demand (BOD) values which are so expensive to deal with.

4.3.1 PENTOSAN

Pentosans, mostly arabinosans – though xylose is also found – occur at significant levels in wheat flour (about 2%) and in soy meal (about 7%) and thus derive from the two most important agricultural commodities in international trade. They have been regarded as important to the baking process in flour, and a variety of pentosanases have been investigated as possible process moderators.

4.3.2 LENTINANS

Lentinans are β(1→3) beta glucans with β(1→6) branching and are found in fungi, particularly basidiomycetes. They have attracted interest since they are said to have anti-tumour activity via stimulation of lymphocytes and macrophages to produce interleukins (Ma et al., 2014).

4.4 FRUCTANS: INULIN AND LEVAN

Fructans are important for two reasons. The first is that they are a potential source of fructose, for which there has been a growing demand as a sweetener. The second is that their biosynthetic mechanisms are different from most of the polysaccharides and may offer the best opportunity for manipulation. In addition, they occur very widely in commercial plants such as cereals and must be considered to be components of raw materials which may affect useful properties. They always contain at least one glucose residue but are otherwise made up of fructofuranose five-membered rings which lead to rather different polymer behaviour as compared with the pyranose ring-based glucans. They are the only polysaccharides based on the furanose structure.

4.4.1 OCCURRENCE AND USES

Fructans (which used to be called fructosans, and are described as 'inulin' in commercial practice) are found in large amounts in only a small number of plants, but are widely distributed. Both monocotyledons and dicotyledons have them. The reason for this variation in content is unknown, and while their primary function is clearly as a reserve carbohydrate, their advantage in comparison with glucans may be connected with chill resistance. They tend to occur instead of glucans such as starch, but many species contain both as well as mannans. Unlike starch, they are mostly localized in vacuoles, and probably in solution, with the larger species possibly present as colloidal sols. This means that they exert a far greater osmotic effect than starch and to some extent break the 'rule' that storage materials should be insoluble.

Fructans were first isolated from *Inula helenium*, hence the name *inulin*, and many other *Compositae* have since been found to make them. They occur in all parts of the plant, notably, from the biotechnology point of view, in the seeds of the Gramineae though only to the extent of 1% or

2% (Meier and Reid, 1982). The best-known cultivated dicotyledon is the Jerusalem artichoke (*Helianthus tuberosus*), the tubers of which, together with cereals, are the major source of inulin used as a special food for diabetics. In practice, the fructose syrups isolated from such sources by hydrolysis isomerizes and contains up to 20% glucose.

With monocotyledons, in the grasses there is a suggestion that fructans and starches are mutually exclusive, with fructans found in temperate species and starches in the tropical varieties. They are widely distributed, and in commercial crops occur in wheat, barley and asparagus, as well as being the major storage carbohydrate of garlic (*Allium sativum*). In wheat, fructans appear to be synthesized before starch during seed maturation.

Apart from use in diabetic diets, some herbal extracts which are probably high in inulin have been advocated as diuretics, while inulin itself is used in kidney clearance tests since it remains intact in the blood after injection for long periods.

Although there is now a substantial market for fructose as a sweetener it is almost entirely made from maize (corn) starch by hydrolysis and isomerization of the resulting glucose. This is a matter entirely of economics and since corn starch is a very cheap and plentiful raw material it is unlikely to be displaced from its major markets. Nevertheless, there are smaller markets in various parts of the world where fructans may be the preferred source of fructose. There are clearly possibilities of manipulating potential crop plants to improve the efficiency of production of fructans. Fructans have been found in the marine algae *Cladophora* and *Rhizoclonium*, but are not present in other species. Starch is also present. Fructans have not as yet been described in fungi. There are also microbial sucrose-dependent fructosyl transferases which may mean that fructans are present in microbes.

4.4.2 BIOSYNTHESIS

Fructans are synthesized by the transfer of a fructose residue from sucrose onto another sucrose receptor molecule (Pollock and Chatterton, 1988). Sucrose is a 'high energy' compound – the ΔG of hydrolysis is -27.6 kJ mol^{-1} compared with UDP (31.8 kJ mol^{-1}) and phosphoryl derivative values of 20.8 kJ mol^{-1} – and it can thus act widely as a donor in polysaccharide biosynthesis. In this instance, although UDP-fructose has been found it is not involved in fructan synthesis. Enzyme systems capable of doing this in three different ways are known to exist leading to the main three types of fructans:

1. The isokestose series – linear chains with $\beta(1 \rightarrow 2)$ links of the general form

$$\text{Glc-}(1 \rightarrow 2)\text{-Fru-}1 \rightarrow (2\text{-Fru-}1)_n \rightarrow 2\text{-Fru}$$

   with a maximum degree of polymerization (DP) of about 35. These are sometimes referred to as *inulins*, but commercial 'inulins' are likely to be a mixture of all types of fructans depending on their source.

2. The neokestose series – in which the glucose residue is linked directly, through the 2 and 6 positions, to two fructose residues and thus the glucose residue is not terminal:

$$\text{Fru-2}{\rightarrow}(1\text{-Fru-2})_m{\rightarrow}1\text{-Fru-}(2{\rightarrow}6)\text{-Glc-}(1{\rightarrow}2)\text{-Fru-1}{\rightarrow}(2\text{-Fru-1})_n{\rightarrow}2\text{-Fru}$$

3. The kestose series – where the linkages are $\beta(2{\rightarrow}6)$ bonds with a general formula

$$\text{Glc-}(1{\rightarrow}2)\text{-Fru-6}{\rightarrow}(2\text{-Fru-6})_n{\rightarrow}2\text{-Fru}$$

Fructans of this type are sometimes called levans, or phleins since they are found in the *Pooideae* grasses. There have also been reports of *branched fructans* though they have not been fully characterized.

The initial synthetic step in all cases is a condensation:

$$\text{Glc-}(1{\rightarrow}2)\text{-Fru} + \text{Glc-}(1{\rightarrow}2)\text{-Fru} = \text{Glc-}(1{\rightarrow}2)\text{-Fru-}(1{\rightarrow}2)\text{-Fru} + \text{Glc}$$

Sucrose   Isokestose

by the enzyme sucrose sucrose-fructosyl transferase (SST) which is active even where the fructans are neokestose based or are levans. The enzyme has been purified to homogeneity and is capable of using some receptors other than sucrose, though it never uses isokestose. It is clearly one point at which site-directed mutagenesis might be used to vary specificity and increase the range of fructans synthesized. Two further enzymes are required to form the fructans. A fructan fructan-fructosyl transferase can transfer fructose residues from one fructan to another, including isokestose as either donor or acceptor. The first of these leads to the formation of $(1{\rightarrow}2)$ links, and thus the inulin or isokestose series, while the second can form the Fru-$(2{\rightarrow}6)$-Glc links found in the neokestose series. These two enzymes have been purified from *Asparagus*. The enzymes responsible for the levan series have not as yet been isolated, though their activity has been demonstrated in cell-free extracts with sucrose as substrate.

A fructosyl transferase from *B. subtilis* has been successfully transferred to tobacco plants where it causes accumulation of fructans up to 8% of the dry weight. This is clearly a development of considerable potential significance for fructose production. These enzymes are attractive as subjects for biotechnology simply because sucrose is readily available in large amounts. Sucrose-dependent transferases which can transfer the glucosyl residue of sucrose are also known (Rastall and Bucke, 1992) and a combination of these with the fructosyl transferases could clearly lead to novel polysaccharides. The use of sucrose as a donor is all the more attractive since the other possibilities involving nucleotide derivatives would certainly be much more difficult. The question is further discussed in Chapter 6.

Cyclic fructans have not been described but would be of considerable interest as a contrast with cyclodextrins.

TABLE 4.4   Fundamental Properties of Jerusalem Artichoke Inulin

Property	
Solubility	Soluble in dimethyl sulphoxide (DMSO)
Molecular weight	3400 Da
Intrinsic viscosity [η]	9 mL g^{-1}
Sedimentation coefficient ($s_{20, w}$)	~0.4 S

Source: From Aziz, B.H. et al., *Carbohydrate Polymers*, 38, 231–234, 1999.

4.4.3 DEGRADATION

Fructan hydrolases have been described and appear to function by removing terminal fructose residues until sucrose remains. There is no evidence of internal group hydrolysis. It is likely that a separate enzyme is involved in splitting the (2→6) links in neokestose and one specific for Fru-(2→6)-Fru links in the levan series has been described. Since acid hydrolysis is easy it may be the preferred method rather than any use of enzymes in process applications. It is also worth noting that fructan hydrolases are absent from the human gut, though they may be present in the microbial flora.

4.4.4 CHARACTERIZATION

Most structural determinations on fructans have been done by methylation. Some solution properties have been measured (Harding et al., 2015), and most analytical methods now depend on gel-filtration/size-exclusion chromatography. The molecular weights tend to be < 50 x 10^3 Da, and there is always some polydispersity based on a single fructose residue difference. This is what would be expected from the synthetic pathway. Table 4.4 summarizes some solution properties (in dimethyl sulphoxide) of Jerusalem artichoke inulin.

FURTHER READING

Halley, P. and Averous, L. (eds) (2014) *Starch Polymers: From Genetic Engineering to Green Applications*, Amsterdam: Elsevier.

SPECIFIC REFERENCES

Akhavan, T. and Anderson, G.H. (2007) Effects of glucose-to-fructose ratios in solutions on subjective satiety, food intake, and satiety hormones in young men, *American Journal of Clinical Nutrition*, 86, 1354–1363.

Aziz, B.H., Chin, B., Deacon, M.P., Harding, S.E. and Pavlov, G.M. (1999) Size and shape of inulin in dimethyl sulphoxide solution, *Carbohydrate Polymers*, 38, 231–234.

Cowan. (1991). *Biotechnology: The Science and the Business*, p. 324, Lausanne, Switzerland: Cordon & Breach Publishers.

Fishman, M.L., Cooke, P., White, B. and Damert, W. (1995) Size distributions of amylose and amylopectin solubilized from corn starch granules, *Carbohydrate Polymers*, 26, 245–253.

Harding, S.E., Adams, G.G. and Gillis, R.B. (2016) Molecular weight analysis of starches: Which technique? *Starch/Stärke*, 68, 1–8.

Harding, S.E., Adams, G.G., Almutairi, F., Alzahrani, Q., Erten, T., Kök, M.S. and Gillis, R.B. (2015) Ultracentrifuge methods for the analysis of polysaccharides, glycoconjugates, and lignins, *Methods in Enzymology*, 562, 391–439.

Henis, Y., Yaron, R., R., Rishpon, J., Sahar, E. and Katchalski-Katzir, E. (1988) Mobility of enzymes on insoluble substrates: The β-amylase starch gel system, *Biopolymers*, 27, 123–138.

Jaeger, K.E. and Eggert, T. (2002) Lipases for biotechnology, *Current Opinions in Biotechnology*, 13, 390–397.

Ma, J., Mo, H., Chen, Y., Ding, D. and Hu, L. (2014) Inhibition of aflatoxin synthesis in Aspergillus flavus by three modified lentinans. *International Journal of Molecular Science*, 15, 3860–3870.

Mathews, C.K. and van Holde, K.E. (1990) *Biochemistry*, pp. 283–284. Redwood City, CA: Benjamin/Cummings.

Meier, H. and Reld, J.S.G. (1982) Reserve polysaccharides other than starch in higher plants, in Loewus, F.A. and Tanner, W. (eds) *Plant Carbohydrates*, Chap. 11, New York, NY: Springer-Verlag.

Morris, G.A., Ang, S., Hill, S.E., Lewis, S., Shäfer, B., Nobbmann, U. and Harding, S.E. (2008) Molar mass and solution conformation of branched α(1→4), α(1→6) glucans. Part I: Glycogens in water, *Carbohydrate Polymers*, 71, 101–108.

Pollock, C.J. and Chatterton, N.J. (1988) Fructans, in Stumpf, P.K. and Cohn, E.E. (eds) *The Biochemistry of Plants. A Comprehensive Treatise*, Vol. 14, Chap. 4, New York, NY: Academic Press.

Rastall, R.A. and Bucke, C. (1992) Enzymatic synthesis of oligosaccharides, in Tombs, M.P. (ed) *Biotechnology and Genetic Engineering Reviews*, Vol. 10, pp. 253–282, Andover: Intercept.

Svensson, B. (1994) Protein engineering in the α-amylase family: Catalytic mechanism, substrate specificity, and stability, *Plant Molecular Biology*, 25, 141–57, 1994.

Tombs, M.P. (1985) Stability of enzymes, *Journal of Applied Biochemistry*, 7, 3–24.

Tombs, M.P. (1990) *Biotechnology in the Food Industry*, Chichester: Open University Press – John Wiley and Sons.

Woods, L.F.J. and Swinton, S.J. (1995) Enzymes in the starch and sugar industries, in Tucker, G.A. and Woods, L.F.J. (eds) *Enzymes in Food Processing*, 2nd Edition, Chap. 8, Glasgow: Blackie.

Zeeman, S.C., Kossmann, J. and Smith, A.M. (2010) Starch: Its metabolism, evolution, and biotechnological modification in plants, *Annual Review of Plant Biology*, 61, 209–234.

Marine Polysaccharides

5

This chapter considers a group of polysaccharides from one of the world's most abundant natural resources: seaweed. These are alginates, carrageenans – the principal ones being kappa (κ), iota (ι) and lambda (λ) – and also agar and its agarose constituent. Another group of abundantly available marine polysaccharides (from the shells of crabs, lobsters and shrimps) is considered: chitin and its soluble derivatives, in particular its partially or fully deacetylated derivative chitosan. Whereas with the storage polysaccharides, the main biotechnological interest has focussed around enzyme modification for enhanced food use, with the marine polysaccharides there has been increasing interest in the application of these molecules in pharmaceutical biotechnology for drug targeting and controlled release. Alginates and carrageenans in particular continue to be important in food technology, most notably as gelling and thickening agents and as stabilizers. In the case of κ-carrageenan, the enhancement of these properties by use in mixed systems with the mannan gums is of current interest.

Marine polysaccharides provide a very exciting area of biotechnological exploitation at the present time and this is particularly true for alginates with some revolutionary developments in microencapsulation technology, including insulin-producing cells for the treatment of diabetics and developments in the use of chitosans as bioadhesive systems in drug delivery including the enhancement of epithelial uptake.

Various interesting reviews can be found dealing with different aspects of industrial uses of all the marine polysaccharides that are discussed in this chapter, and the student can find these in the extended Further Reading at the end. Of particular interest - as CO_2 fixation is an issue- is the review by Dibenedetto (2011) on the potential of aquatic biomass for fixation and energy production.

5.1 ALGINATES AND BROWN SEAWEED

Seaweed has been used as part of the human diet for at least 3000 years and the Vikings used it for wound dressing. It is no surprise that since the English chemist E. C. C. Stanford first obtained it from seaweed in 1886, isolated alginate – the most interesting structural polysaccharide of brown seaweed (*Phaeophyceae*) – has found, particularly in the last 60 years, use in the food, biomedical, pharmaceutical and other bioindustrial areas.

After Stanford's discovery, W. L. Nelson and L. Cretcher in 1929 showed it to be a polyuronic acid, and subsequent studies (Fischer and Dörfel, 1955) showed two types of uronic acid present: β-L-mannuronic acid (ManA or further abbreviated as 'M') and α-D-guluronic acid (GulA or 'G') (Figure 5.1) distributed as blocks of GulA, ManA and heteropolymeric mixed sequences (GulA-ManA, usually alternating). Industrialists often refer to them by their 'M:G' ratio (not to be confused with the G:M ratio for galactomannans [Chapter 2] which refers to different residues), and this ratio has an important bearing on alginate properties as we shall see in the following.

Except under low pH conditions (pH <3.5), both poly-acids are dissociated and the residues are referred to as guluronate and mannuronate, respectively: the pK_a of D-mannuronic acid is 3.38 and that of L-guluronic acid 3.65. In -M-M-M- sequences residues are linked β(1→4); in -G-G-G-, α(1→4); in -M-G-M-G- there is corresponding alternation from β(1→4) to α(1→4). M and G differ in the position of the COO⁻ group and the three-dimensional (3D) preferred ring structure, with M in the $^4C_1$ and G in the $^1C_4$ chair conformations (see Figure 1.3): these features are critical in governing the structure of poly-G, poly-M and poly-MG chains, with for example poly-G exhibiting a buckled ribbon-like structure and poly-M a somewhat more flexible extended ribbon-like structure (Figure 5.2).

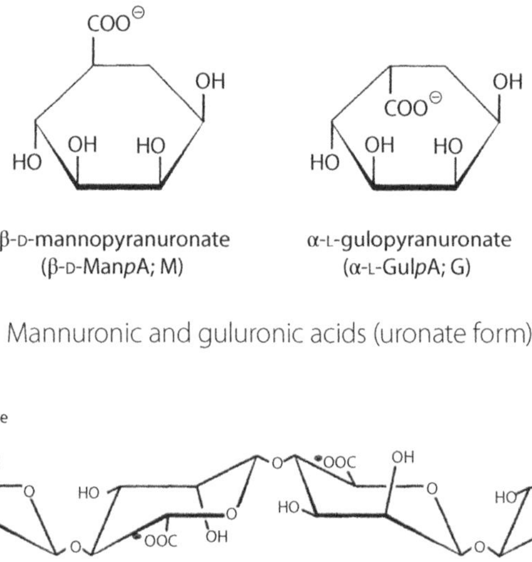

FIGURE 5.1   Mannuronic and guluronic acids (uronate form).

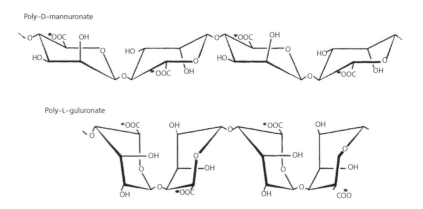

FIGURE 5.2   Poly-G and poly-M alginate sequences give different structures.

The stiffness of chains increases in the order poly-MG→poly-M→poly-G. Pure poly-M alginates are not found in seaweed species, but are produced by some bacteria: the latter are the subject of considerable attention for the design, with the assistance of epimerases, for converting M to G residues of 'purpose built' alginates (Ertesvag et al., 1996): this will be considered in Chapter 6 on bacterial polysaccharides. It remains, of course, to be seen if epimerases isolated from bacteria can be commercially applied *in vitro* to provide useful adjustment to the properties of seaweed alginates.

The average degree of polymerization of an alginate varies between seaweed species from ~80 to ~1000 giving molecular weights in the range from ~20,000 to over 200,000 daltons, with bacterial alginates somewhat smaller.

5.1.1 EXTRACTION AND PRODUCTION

There are currently four main types of brown seaweed that provide commercial alginate (Table 5.1), with annual production of approximately 30,000 tonnes per annum. Although the global resource of brown seaweed is approximately 10 times that of current demand, because the majority are in remote areas where transport costs are excessive there is actually a lack of raw material; this problem is accentuated in that the alginates of most commercial value (i.e. those with low M:G ratios) tend to be located in these areas. It seems that younger seaweed is rich in poly-M, particularly the reproductive bodies of *Ascophyllum nodosum* and *Fucales* (nearly pure poly-M). As a seaweed ages, the proportion of guluronate increases due to the action of *epimerases* (Figure 5.3). Other brown seaweeds used for alginate production come from *Laminaria japonica* and also *Ecklonia, Durvillea, Fucus, Egregaria, Nereocystis, Pelagophycus, Undaria* and *Lessonia* spp. Brown seaweeds also produce other polysaccharides such as laminarin, although their biotechnological importance is relatively small compared to alginate.

The alginate content of a seaweed varies from around 10% to 50% and depends not only on the species but also on the time of year and even varies within the plant itself, with the frond (leaves), stipe (stem) and holdfast (anchor) showing progressively more poly-G. This feature correlates well with biological function, with greater rigidity required in the region anchoring the seaweed to a rock, whereas greater flexibility (more poly-MG and poly-M) is needed in the wave-bashed fronds which flap about near the surface of the water. Alginates are extracted from the seaweed

TABLE 5.1   Chief Brown Seaweed Sources of Alginate

Source	Alginate (%)	M:G Ratio
Laminaria hyperborea (stipes)	25–38	0.5
Laminaria digitata	18–26	1.5
Macrocystis pyrifera	20–31	1.6
Ascophyllum nodosum	19–30	1.9

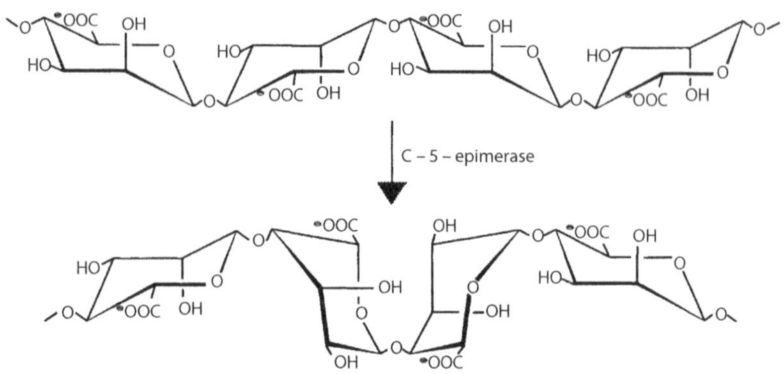

FIGURE 5.3   Action of epimerases. (Adapted from Skjåk-Bræk, G. and Espevik, T., *Carbohydrates in Europe*, 14, 19–25, 1996. Reprinted with permission.)

by milling and washing followed by dissolution by heating with alkali and precipitation with calcium chloride. This is followed by decolouration and acid treatment and soaking in 3% sodium carbonate solution. The process is finished by drying and milling to produce a white or off-white powder. After isolation, commercial operators often blend different sources of alginates (with different M:G ratios) to give the desired properties: an example is the Nutrasweet-Kelco (San Diego, United States) 'Manucol DM'.

5.1.2 PROPERTIES

The fundamental properties of a typical commercial ('high-G') alginate are summarized in Table 5.2. Purity of commercial samples varies from source to source. Protein content is usually a fraction of a per cent as assessed by Kjeldahl analysis of its nitrogen content. Nuclear magnetic resonance measurements on oligosaccharide fragments can be used to assess the microheterogeneity of a sample, including the presence of other sugar residues. In particular, use of specific lyases (poly-β-D-mannuronolyases and poly-α-L-guluronolyases) to produce saccharide and oligosaccharide fragments followed by subsequent analysis can be used to assess the M:G ratio and the approximate fraction of poly-M, poly-G and poly-MG blocks. NMR studies have shown that the poly-G blocks are often punctuated by isolated M residues and that poly-MG sequences are terminated by MMG sequences.

In terms of molecular weight averages and distributions, commercial alginate preparations have been shown by absolute light scattering and analytical ultracentrifugation techniques to have molecular weights (weight averages, M_w) in the range 200–500 kDa.

Alginates have very high thermodynamic non-idealities: for example A_2M_w is about 1000 mL g^{-1} for *Laminaria hyperborea* alginate in a buffered solution of ionic strength $I = 0.3$ (Horton et al., 1991b). In practical terms, this means that if you measure the apparent molecular weight at a single concentration of 0.2 mg mL^{-1} you underestimate the true molecular weight (see Section 1.1.7) by a factor $\sim(1 + 2A_2M_wC) = (1 + 2 \times 1000 \times 0.0002) = 1.4$

TABLE 5.2   Fundamental Properties of an Alginate from
Laminaria hyperborea

Property	Characteristic
Polymer type	Polyanionic; block copolymer
Conformation type	Zone B/C (flexible rod/stiff coil)
Wales–van Holde ratio $k_s/[\eta]$	0.2–0.5
Mark–Houwink 'a' coefficient	~1.0
M:G ratio	0.5–0.6
Solubility	Depends on M:G ratio and temperature
Molecular weight	~230 kDa
Gelation properties	Needs bivalent ions (usually Ca^{2+})
Potential sites for biomodification	M residues (conversion to G residues by epimerases); esterification of COO^- groups: propylene glycol alginate

Sources: Data from Horton, J.C. et al., *Biochemical Society Transactions*, 19, 510–511, 1991; Horton, J.C. et al., *Food Hydrocolloids*, 5, 125–127, 1991.

or 40%. This is a feature of their enormous exclusion volumes, which in turn is a manifestation of the great ability of these molecules to immobilize water. These high exclusion volumes are also reflected in their high intrinsic viscosities ($[\eta]$ which range from ~700 to 1200 mL g^{-1}). The $[\eta]$ versus M_w Mark–Houwink (MHKS) exponent a is approximately unity (with some variation with ionic strength) suggesting a semi-flexible coil model (Conformation Zone C) in an aqueous environment. Alginates with a large proportion of alternating MG blocks have the most flexibility as noted earlier (a ~0.7) whereas those with a large proportion of extended G blocks have the largest rigidity (a ~1.3), approaching Conformation Zone B. Similar conformations are observed under the electron microscope. Alginates gel in the presence of bivalent cations, notably Ca^{2+}. The gel strength depends on the molecular weight distribution and on the M:G ratio (with smaller ratios giving stronger, but less flexible gels): this in turn is directly related to (1) the flexibility of the polymer chain, as noted earlier and (2) the ability of the chain to bind divalent cations. Alginate gelation is one of a number of important properties that can potentially be enhanced by tailoring the alginate by using enzymes or other methods to achieve the specific type of property required.

5.1.3 ENZYMATIC AND CHEMICAL MODIFICATION

There are two potential types of enzyme that could be used commercially to modify alginate properties. The first are mannuronan epimerases which convert mannuronic acid to guluronic acid (Figure 5.3). *In vivo* they are endo – acting only on polymer blocks of at least 10 residues and containing at least two consecutive M residues. There is particular interest in bacterial alginate epimerases and the biotechnology behind these will be considered in Chapter 6. The other main type of enzyme is lyases that are specific for either β-D-mannuronate or poly-α-L-guluronate and

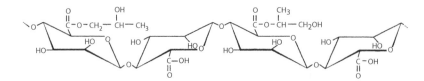

FIGURE 5.4   Propylene glycol alginate.

which, if allowed to act until completion, degrade the alginate to a digest of oligosaccharide units: lyases, together with associated NMR measurements are therefore of particular use for probing the structure and the M–G composition and arrangement of an alginate.

The M:G ratio can also be increased by chemical methods (Skjåk-Bræk et al., 1996): application of CO_2 at high pressure (>150 atm, where 1 atm = 10^5 Pa) to solid alginate, but again there have been no commercial developments.

Various other attempts have been made to modify alginate properties: these include amide, amine and ester derivatives, the most important being propylene glycol alginate (PGA) (Figure 5.4), made by reacting propylene glycol with the COO^- groups on the alginate to partial esterification (40%–85%). This is particularly useful under high-acid conditions, where alginate and other hydrocolloids become unstable or insoluble. Salad dressing is a good example: high viscosity at very low shear rates gives good emulsification with long shelf life without sedimentation or phase separation; lower viscosity at higher shear rates gives on the other hand good pourability. The bulk of biotechnological interest is, however, in terms of unsubstituted alginates.

Yang et al. (2012) and Pawar and Edgar (2012) have recently reviewed the chemistry, properties and applications of various derivatives of alginates that can be useful both for the physical and biological activities seen from studies on alginates.

5.1.4 USES OF ALGINATES

In the plant cell wall, alginates occur in fibrous form and indeed alginate fibres were once considered for commercial use as a cellulose substitute, although the main uses of alginate fibres (or Ca^{2+} and Ca^{2+}/Na^+ salts) are now in surgical cuts and dressings. In dentistry, alginates along with other marine polysaccharides are used in the manufacture of dental impression formulations with a range of properties: high poly-G alginates provide firm impression media whereas high poly-M or poly-MG alginates give softer 'pudding-like' impression materials. The main uses of alginates are as follows:

- Gelling and thickening agents and emulsion stabilizers in the food industry
- Thickening agents for use in the textile and paper industries
- Wound healing
- Microencapsulation and controlled release systems

More detailed considerations of the commercial usefulness of alginates, including biomedical applications, can be found in Onsoyen (1996), Rinaudo (2006), D'Ayala et al. (2008), Laurienzo (2010), Lee and Mooney (2012).

5.1.5 GELLING/THICKENING/EMULSIFIERS/FIBRES IN THE FOOD INDUSTRY

5.1.5.1 GELATION

In the controlled presence of calcium ions, alginate will gel via a process commonly known as the *egg-box* process (Figure 5.5), a process which appears to apply also to low methoxyl (DE < 50%) pectins. In the case of alginates, this egg-box process involves intrachain poly-G regions with the optimum calcium ion concentration (usually up to 0.5 M) depending on the proportion of poly-G in the alginate. Table 5.3 compares the performance of two commercial alginates of different M:G composition; higher poly-G alginates at the appropriate Ca^{2+} concentration give the stronger gels, higher poly-M are less sensitive to Ca^{2+} ion concentration.

The calcium ion concentration – and the speed with which the calcium ions are added – can be critical: too little gives rise to weaker gels whereas too much added too quickly tends to give rise to precipitate formation. Considerable attention has been paid by the industry to the control of the gelation process using for example 'diffusion strips' or so-called 'onion rings'. Further control can be obtained by the use of sequestrants. Sequestrants, which may also be used to remove water hardness to assist alginate dissolution, such as sodium hexametaphosphate, tetrasodium pyrophosphate, disodium phosphate and sodium citrate, remove any residual calcium left after the manufacturing process and compete with added calcium during gel preparation. Recent research (Padol et al., 2015) has shown that the addition of small chain oligomers of guluronate can also lead to more uniform gels.

Alginate gels are generally thermo-irreversible, that is they will not remelt on heating. They also give poor stability at lower pHs. Stable mixed

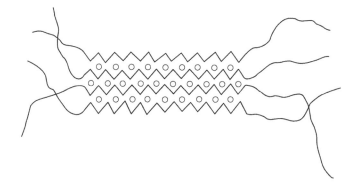

FIGURE 5.5 Egg-box mechanism for the formation of alginate gels. (Reprinted from Grant, G.T. et al., Biological interactions between polysaccharides and divalent cations: The egg-box model, *FEBS Letters*, 32, 195–198, 1973, with permission from Elsevier.)

TABLE 5.3   Effect of Calcium Ions on Commercial Alginates of Different M:G Ratio. The lower the M:G, the 'stronger' the gel. Calcium ion concentration is represented as the proportion of alginate in the calcium salt form as opposed to the monovalent salt form. The number of "+" reflects the relative magnitude of the effect.

	Manucol DM		**Manugel DMB**	
	~60%	~100%	~60%	~100%
Deformation	+ + +	+ +	+ +	+
Gel strength	+	+ +	+ + +	+ + + +
Elasticity	+ +	+ +		
Brittleness			+ +	+ +
Syneresis	+	+	+ +	+ + +

Source: Data from Nutrasweet-Kelco Ltd.

alginate–(high methoxyl) pectin gels, however, can be formed at low pH (<3.8) and are thermo-reversible. The stability at low pH makes them attractive for use in low-calorie jams and jellies.

5.1.5.2 THICKENING AND SURFACE ACTIVE PROPERTIES

The high viscosity and hydration properties of alginates make them useful in salad dressings which usually contain over 35% of oil: use of an alginate thickener gives the dressing acceptable flow characteristics and prolongs the shelf life. Propylene glycol alginate, PGA (Figure 5.4) is particularly useful because of its acid stability and since the hydrophobic propylene glycol group confers on the alginate additional emulsification stabilization properties. The surface active properties of PGA are also used to good effect in stabilizing the foamy head of beers.

5.1.5.3 FIBRES AND FILMS

Controlled drying can be used to make fibres or films of alginate which can be used to protect fresh meat and fish from surface bacterial contamination and dehydration. These films have potential use in non-solvent-based sealant systems for cans, with the alginate sealing compound applied as a thin strip into the slit of the can: the efficient administration of the film requires alginate with the appropriate rheological properties. Another important use is for ceramics, where their water holding capacity slows down surface drying, thus reducing problems due to surface cracking and distortion prior to the drying and baking process.

5.1.6 THICKENING AGENTS FOR USE IN THE TEXTILE AND PAPER INDUSTRIES

Alginates find their biggest commercial use in the textile industry where they provide an ideal medium for the application of printing

FIGURE 5.6   Textile printing: Addition of low-viscosity alginate to the dye helps in the printing of fine lines. (Reprinted from Onsoyen, E., *Carbohydrates in Europe*, 14, 26–31, 1996. With permission.)

dyes. The high viscosity results in slow spreading of the dye – important for fine-line printing – whereas its chemical inertness to the dye (due to the absence of primary hydroxyl groups) under alkaline conditions means that very little dye is wasted when the alginate paste is washed off. In a similar way, the addition of alginates to clay pastes used in coating paper thickens the paste and prevents agglomeration of the coating. Higher print quality also results because of the inertness of the alginate to printing ink. For the printing of fine lines (Figure 5.6), the optimum results are obtained using a low-viscosity alginate preparation. The inertness of alginate films to oil and grease also reduces the problem of staining.

5.1.7 WOUND HEALING

The *Vikings* reputedly used brown seaweed in wound management. As often happens before the world had the power of chemical insight, much progress had been made by simple trial and error of known materials, although it has taken almost a millennium to show that these old warriors had made an inspired choice. The modern introduction of alginates into wound management followed the commercial production in the 1930s of calcium alginate fibre. It has only been relatively recently the true importance of these materials has been realized (Schmidt, 1991; Wang et al., 2015). This corresponded to an appreciation that wounds heal faster if kept in a moist environment with sterile exudate rather than being allowed to dry. The ability of calcium alginate fibres and films to retain a moist environment, whilst providing haemostatic activity, resulted in various commercial wound healing products becoming available – such as *Sorbsan*, *Kaltostat*, *Kaltoclude*, *Tegagel*, *Stop Hemo* and *Ultraplast*. It is believed that the calcium ions themselves assisted with the blood clotting process.

5.1.8 MICROENCAPSULATION AND CONTROLLED RELEASE SYSTEMS

The use of alginate (particularly high-G) gelled encapsulation and release systems has a large literature and includes the immobilization and controlled release of pesticides, biocatalysts and drugs (see e.g. Skaugrud, 1993; Schacht et al., 1993; Thu et al., 1996b). They can also be used as a protective encapsulation and immunoprotecting medium for the immobilization of whole cells, which can in principle then be targeted so as to release their enzymes and hormones at the desired site of action (Table 5.4).

When placed in an aqueous environment, calcium alginate gels will release the enclosed active agent at a rate determined by the solubility of this agent in the aqueous medium. Surface coating of alginate beads used for the immobilization with a cross-linked alginate/polycationic

---

TABLE 5.4   Examples of the Use of Alginate-Encapsulated Immobilized Cells

System	Product
Bacteria	
Erwinia rhapontici	Isomaltose
Pseudomonas denitrificans	Drinking water
Zymonas mobilis	Ethanol
Fungi	
Penicillium chrysogenum	Penicillin
Kluyveromyces bulgaricus	Hydrolysis of whey
Claviceps purpurea	Alkaloids and ergotoxins
Saccharomyces cerevisiae	Ethanol
Saccharomyces bajanus	Champagne
Brown algae	
Botryococcus barunii	Hydrocarbons
Blue–green algae	
Anabena spp.	Ammonia
Plant cells	
Chatharanthus roseus	Alkaloids
Daucus carota	Alkaloids
Various plants	Artificial seeds
Plant protoplasts	Cell handling, microscopy
Mammalian cells	
Hybridoma	Monoclonal antibodies
Islet of Langerhans	Insulin/implantation
Fibroblasts	Interferon-β
Lymphona cells	Interferon-α
Hepatocytes	De-esterifiers of blood
Neural cells	Neurotransmitters and trophic factors

Sources: Data from Blinden, G. and Guiery, M. (eds): *Seaweed Resources in Europe—Use and Potential.* pp. 221–257. 1991. Copyright Wiley-VCH Verlag GmbH & Co. KGaA. Reprinted from Skjåk-Bræk, G. and Espevik, T., *Carbohydrates in Europe*, 14, 19–25, 1996, and references cited therein, with permission.

polymer such as poly-L-lysine stabilizes the bead and helps to control the porosity of the release mechanism. Figure 5.7 attempts to give an overview of the microencapsulation process, showing a summary of the basic properties of alginates which make them excellent immobilization materials (Figure 5.7a), an outline protocol for the immobilization of insulin producing cells involving the application of calcium ions to the alginate in the presence of the material being immobilized (Figure 5.7b) and a schematic structure of an alginate capsule encapsulating such cells, with an additional polycationic layer (Figure 5.7c). In this case, polylysine was used, although chitosans have also been considered. Bead sizes can be produced ranging in size from a few microns to several millimetres in diameter. As with other applications, the particular alginate can be chosen depending on the application, with high poly-G alginates appearing to give stronger beads with high resistance to anti-gel agents such as Na^+ and Mg^{2+} ions or sequesters of Ca^{2+} ions, while high poly-M alginates give more elastic beads (Thu et al., 1996a). One of the biggest advantages of a microencapsulation system is that it can provide fragile cells with a non-destructive barrier against immune attack. Of particular value has been the incorporation of insulin producing cells (islets of Langerhans) for treatment of diabetics (Richardson et al., 2014).

The benefit of including a polycation at the surface of the bead has also been illustrated with poly (ethylene imine) and Figure 5.8 demonstrates control of the release of three pesticides (dichlobenil, propanil and carbofuran). Chitosans have also been considered (see Section 5.4.4). For drug delivery using this type of release mechanism and other biomedical applications, particularly intravenous systems, alginate preparations of the highest purity are required, and 'ultrapure grades' of alginate are commercially available which meet these requirements: these include filtration through a 0.22-μm filter, negligible amounts of endotoxin and the optimum M:G ratio to minimize undesirable immunological response.

The microcapsules based on alginates have also been used in the treatment of chemically induced renal failure in rodents with success and it is suggested that this could be used instead of conventional dialysis (O'Loughlin et al., 2004). A general overview of alginates for drug delivery has been reviewed recently by Cardoso et al. (2016).

5.1.9 BIOLOGICAL ACTIVITIES OF ALGINATES

Ueno and Oda (2014) have given an overview of the macrophage stimulating activities of alginates as well as their anti-oxidant activities. They discuss the effects depending on the source from which the alginate has been produced, the effect related to the M:G ratio as well as the effect depending on the molecular size. The target for their measurements is the production of TNF-α by the RAW264.7 cells, a mouse macrophage cell line. The results showed that alginates with M_w higher than 30,000 had significant effect on the RAW cells, while those with M_w below 10,000 had no significant activity. Other samples with M_w around 20,000 gave

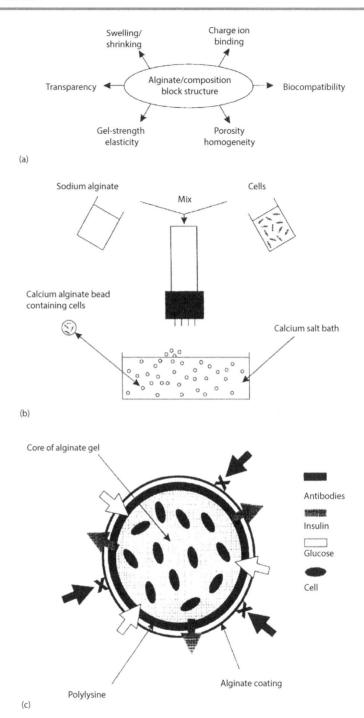

FIGURE 5.7 Microencapsulation systems using alginate. (a) Properties of alginate which make them highly useful as an immobilization material. (b) Schematic protocol for entrapment of cells in calcium alginate. (c) Immunoisolation of insulin-producing cells in an alginate polycation microcapsule. (Reprinted from Skjåk-Bræk, G. and Espevik, T., *Carbohydrates in Europe*, 14, 19–25, 1996. With permission.)

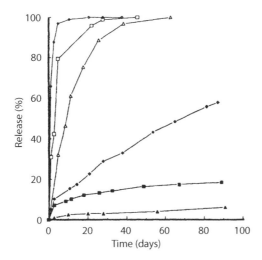

FIGURE 5.8   Comparison of sustained release from alginate bead systems with and without a polycationic coating. Release of dichlobenil (triangle), propanil (square) and carbofuran (diamond) from dried calcium alginate beads (open symbols) and polyethyleneimine (PEI)-treated alginate (closed symbols) granules (treatment: 10% vol. 'Corcat 12' at pH 3 for hours). (Schacht et al. [1993]. Reproduced by permission of the Royal Society of Chemistry.)

significant different activities, but their M:G ratios differed. Those with M:G ratio of approximately 2 had a level of activity 10-fold that of those with M:G ratio of 1.5. The authors conclude that higher M_W and a ratio of M:G higher than 2 is important for a high production of TNF-α. It was also shown that the alginates induced the production of NO in the RAW264.7 cells, and enzymatically depolymerized, unsaturated alginate oligomers had more potent activity that the parent molecule. These authors also suggest that alginates also may have an antioxidant activity, but the mechanism is not understood. Skjåk-Bræk et al. (2000) have also proven that the polyuronides have important effects on the immune system, as exemplified with the alginates.

5.2 CARRAGEENANS AND RED SEAWEED

The carrageenans are a group of related linear and sulphated polymers and are the main structural polysaccharides of red seaweed (*Rhodophyceae*). Three carrageenans are commercially important, iota (ι), kappa (κ) and lambda (λ), and like alginate they are used as thickeners and gelling agents in the food industry. They are also widely used in toothpaste manufacture, although they have not been exploited to the same extent as alginates. Like alginates they have a long history and are named after the coastal town of Carragheen in Ireland where for hundreds of years dried and bleached red seaweed referred to as 'Irish moss' had been used as a food geller and thickener. By the nineteenth century, the same material was being used for beer clarification and in textiles. Again, just like alginates, by the early part of the twentieth century the carrageenan itself was being extracted and used directly.

The structure of these carrageenans is largely of the alternating repeat type (Figure 1.5). The principal gel formers, iota and kappa, have a similar repeat structure of the form

··· →3) β-D-Galp-4-sulphate-(1→4) 3,6-anhydro-α-D-Galp (1→ ···

with the anhydro-galactose residue sulphated at position 2 in iota. Figure 5.9 shows the Haworth projection of the disaccharide repeat for these and the other main commercial carrageenan, λ-carrageenan. It can be seen that lambda is quite different from the other two in having virtually no anhydro-oxygen bridge residues. This has important functional consequences since iota and kappa are widely (but not exclusively) thought to exist in a so-called 'ordered' or helical form through heat-sensitive intrachain hydrogen bonding. Such ordered structures do not exist in lambda which also does not gel. The precise conditions of salt and temperature under which this ordering occurs, its helical nature (whether it is single or double) and the transition to disordered form have been the subject of considerable debate. This is important since it is at the root of our understanding of the mechanism of gel formation and also of commercially useful interactions with other polysaccharides. Similar uncertainties apply to other polysaccharides that under certain conditions have ordered structures, notably agarose and xanthan.

From a biosynthesis point of view (Craigie and Wong, 1979) this is interesting in that the anhydro is also missing during the early stages of synthesis in the Golgi apparatus of the common galactan precursor of

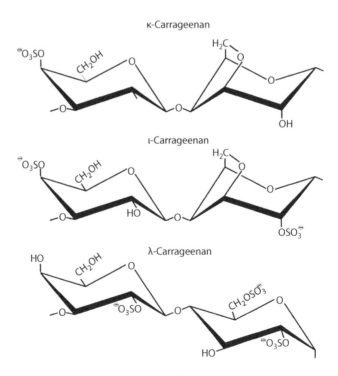

FIGURE 5.9 Carrageenan disaccharide repeat unit.

the iota, kappa and lambda forms. Sulphation is thought to take place at a much later stage and in the cell walls. As with other structured polysaccharides such as cellulose and xanthan, the ordered hydrogen-bonded structures of iota and kappa also make them considerably less soluble than lambda. The occasional presence of a disaccharide pair lacking the anhydro-oxygen bridge in ι- and κ-carrageenan gives rise to structural kinks in the polysaccharide chain that terminates a helix and helps to solubilize the polysaccharide. The presence of these helices in fibres of carrageenan has been visualized directly by x-ray fibre diffraction studies (Lama and Segre, 1980). The helical domains are thought to provide the sites for further aggregation or 'junction zones' and apparently give rise to fibre-like structures of ~8–20 nm in diameter (compared with ~2 nm for a single strand). These junction zones play an important part in both the strength and hysteresis (remelting temperature is greater than the gel setting temperature) of carrageenan gels. Also important in the gel properties is the type of cations present in the surrounding medium, with for example K^+ giving more useful gels than Na^+: sodium salts of carrageenan (particularly κ-) are correspondingly more soluble than potassium salts.

5.2.1 EXTRACTION AND PRODUCTION

The main kind of red seaweed used for carrageenan production is from *Eucheuma* with the main producers being the Philippines and Indonesia (~30,000 tonnes of seaweed per annum). Other important seaweeds used are *Iridaea* and *Gigartina* spp. (~10,000 tonnes per annum from South America) and *Chondrus crispus* (~5000 tonnes per annum from Canada). Table 5.5 summarizes the main types. The carrageenan content of a seaweed depends not only on the species but the time of year and even varies within the plant itself. Sulphate content of κ-carrageenan is ~25%–30% that of ι-carrageenan is on average higher (28%–35%) and λ-carrageenan 32%–39%.

In terms of production, growth tanks have proved uneconomical although 'mariculture' has proved successful in the Philippines – the main producer of the raw material (the dried seaweed) – and Indonesia. Seaweed harvesting is still largely done by hand from the shore, at low tides or from boats using rakes.

Carrageenans are extracted from the seaweed following a procedure similar to that for alginate. The seaweed is washed and treated with strong alkali at high temperature for a day: this digests the weed and also enhances the degree of anhydro-oxygen bridge formation (κ, ι). Lower

---

TABLE 5.5   Chief Red Seaweed Sources of Carrageenan

Source	Main Type of Carrageenan
Eucheuma spp.	κ, ι and κ/ι hybrids
Iridaea spp.	κ/ι hybrids
Gigartina spp.	κ
Chondrus crispus	κ

pHs are used if this is not required (e.g. for non-gelling applications and the extraction of λ). The extraction mixture is then clarified by course filtration followed by centrifugation and fine filtration through porous silica and then usually concentrated in vacuum evaporators to a concentration of ~20–30 g L^{-1}. Because of its poor solubility in the presence of K$^+$ ions, κ-carrageenan can be precipitated as gelled fibres by 1%–5% potassium chloride. Other carrageenans as well as κ can be precipitated by addition of 2-propanol to give a fibrous coagulum. A freeze–thaw followed by centrifugation and pressing is used to assist in the drying process.

5.2.2 PROPERTIES

Table 5.6 summarizes the fundamental properties. The purity of commercial samples can be variable and various grades are produced by the commercial manufacturer depending on the application. Food-grade carrageenans often contain added dextrose or sucrose. Protein impurity, as assessed by sodium dodecyl sulphate (SDS) gel electrophoresis, N-content or ultraviolet (UV) spectroscopy at 280 nm, depends on the grade.

In terms of molecular weights, commercial κ-carrageenan preparations (the most commonly used type) have been shown (by light scattering and analytical ultracentrifugation) to have molecular weights (weight averages, M_w) of the order of 300 kDa (Slootmaekers et al., 1991; Harding et al., 1997), and for λ-carrageenan 300–900 kDa (Almutairi et al, 2014) and are thus somewhat larger than commercial alginates. κ-carrageenans appear though to have smaller exclusion volumes than alginates in solution, as judged from the lower intrinsic viscosity (~650 mL g^{-1} in an 0.1 M ionic strength Na$^+$ buffered salt solution). They have conformations ranging from semi-flexible coils (zone C) depending not only on the charge (i.e. degree of sulphation and pH) but also on the extent of helix

---

TABLE 5.6   Fundamental Properties: Carrageenans

Property	Characteristics
Polymer type	Sulphated polyanion; alternating repeat
Solubility	κ, ι: K$^+$, Na$^+$ salts soluble >60°C; Na$^+$ salts soluble at room temperature; λ: soluble at all temperatures
Conformation	Helical, flexible rod/stiff coil – Zone B/C (ι, κ); stiff coil – Zone C (λ)
Molecular weight	~300 kDa (κ); ~300–900 kDa (λ)
Intrinsic viscosity [η]	~650 mL g^{-1} (κ)
Sedimentation coefficient, $s_{20,w}$	~4.2 S (κ)
Wales–van Holde ratio $k_s/[\eta]$	~0.9 (κ); 1.0 (λ)
Gelation properties	κ, ι: thermoreversible from cooling hot aqueous solutions (>5 g L^{-1}). Strength depends on concentration and salt. λ: no gel
Key sites for biomodification	Anhydro-oxygen bridge; degree of sulphation; type of salt (K$^+$, Na$^+$ etc.); double-helical regions – sites for synergistic interaction with other polysaccharides

structures and anhydro content. λ-carregeenans on the other hand appear to have somewhat higher molecular weights, approaching 1 MDa with higher intrinsic viscosities (Almutairi et al., 2013). They also have a higher charge.

Although as already noted λ-carrageenan does not gel, the κ- and ι-carrageenans form thermoreversible gels above concentrations of 0.5% or 5 g L^{-1} (as low as 2 g L^{-1} in some mixtures of κ- or ι- with proteins, small sugars or other polysaccharides). The gel strength depends on the polysaccharide concentration and type of salt with K$^+$ > Ca^{2+} > Na$^+$. Gel syneresis (weeping or water shedding) problems prevalent in κ-carrageenan gels can be reduced by the presence of impurity or the addition of different polysaccharides, as will be considered in the following section.

Carrageenan products can have short shelf lives because of poor stability, particularly the ι- and κ-forms where the susceptibility to degradation is thought to be related to the intramolecular strain imposed by the anhydro-oxygen bridge. They are particularly susceptible to acid hydrolysis at low pHs (<4) and can be subject to oxidative degradation particularly at greater than ambient temperatures.

5.2.3 BIOMODIFICATION: SYNERGISTIC INTERACTIONS WITH GALACTOMANNANS

The properties and functional use of carrageenans seem to be closely related to the anhydro-galactose content (absent in λ), and in the κ- and ι-forms this can be reduced to some extent by the strength of the alkali used in the extraction procedure. The anhydro content can be increased by enzymes (galactose-6-sulphurylases) that promote essentially the opposite reaction.

The sizes of the carrageenan molecules can be reduced by the use of bacterial enzymes from the species *Pseudomonas carrageenova* which can be obtained with κ-, ι- or λ-specificity: deliberate enzymatic reduction of carrageenan size is, however, no longer of commercial importance, probably because of past concerns of the possible toxicity of low-molecular weight carrageenan.

The main target for enhancement of properties, particularly in terms of gelling and thickening properties, has been the effect of the addition of different polysaccharides: considerable attention in particular has come from the effect of the addition of locust bean gum (the related galactomannan, guar, has little effect), and to a lesser extent xanthan, and evidence for *synergistic interactions* has come from hydrodynamic properties (effect on [η], $s_{20,w}$ etc.) and rheological studies – the effect on storage and elastic moduli. As discussed in Chapter 2, addition of locust bean gum (which does not gel of its own accord) to a solution of κ-carrageenan can reduce the gelling (total polymer) concentration to ~2 g L^{-1} and gives stronger gels with reduced syneresis problems. The mechanism of enhancement has been proposed to be through interactions (presumably hydrogen bond) between the helical regions and unbranched regions of the locust bean gum with the branched regions of the latter assisting the formation of a network (cf. Figure 2.13).

5.2.4 USES OF CARRAGEENANS

Unlike alginates, carrageenan use is dominated by food applications, particularly in connection with milk-based products. The only other bio-applications of commercial significance are in its use in toothpaste and, along with alginates and chitosans, its potential use for microencapsulation and controlled release of drugs and flavours. This area has been well developed and is described later in this chapter.

5.2.5 GELLING/THICKENING IN THE FOOD INDUSTRY: MILK PRODUCTS

As with alginates and pectins, the ability of carrageenans to form gels, retain water and give emulsion stability, together with relatively low production costs, makes carrageenan popular with the food industry. κ- and ι-carrageenans at low concentration (<~5 mg L^{-1}) and below the helix disorder transition temperature (~50°C) are particularly useful for stabilizing dairy products, largely deriving from favourable electrostatic interactions between polyanionic carrageenan (at pHs >~3.5, the pK_a) and the net polycationic (at pHs < pI (~7)) casein and other proteins in milk. Specifically, they

- Improve heat stability
- Inhibit the undesirable action of milk lipases (lipolysis)
- Inhibit coagulation in skim milk
- Inhibit cream and skim separation
- Stabilize calcium sensitive caseins

They are particularly popular with gelled milk deserts. Again because of favourable electrostatic interaction, κ- and ι-carrageenan in milk will give firm gels at ~1.5 g L^{-1} (about five times less than that in the absence of protein). They are also often used in conjunction with other polysaccharides (notably locust bean gum and carboxymethylcellulose) in non-dairy fruit-gelled deserts, allowing stable suspension of chunks of fruit without the necessary addition of large amounts of sucrose, making the application attractive for diet and diabetic products. Table 5.7 summarizes the food applications and the concentrations of carrageenan normally required.

5.2.6 TOOTHPASTE

The high gel melting temperature of ι-carrageenan, together with its thixotropic flow properties and good rinseability, makes it attractive as a toothpaste thickener. Concentrations up to 1% are normally employed. ι-carrageenan is often used together with carboxymethylcellulose where

TABLE 5.7   Carrageenans in Food Products

System	Carrageenan Type and Content (%)[a]
Chocolate milk stabilization	κ, 0.02–0.03
Filled milk (vegetable fat substituted) stabilization	κ or ι, 0.02–0.04
Synthetic (soy, casein, vegetable fat) milk stabilization	λ, ι, 0.05
Evaporated milk stabilization	κ, 0.005
Custards/flans/milk puddings (stable gels)	κ, ι, mixed, 0.2–0.4
Ice cream stabilization	κ, 0.02–0.03
Instant ice cream stabilization	λ, mixed, 0.1
Cottage cheese stabilization	κ, 0.01–0.04
Whipped cream/desert stabilization	κ, 0.03–0.05
Instant pie filling cream thickener	Mixed 0.1–3
Coffee-white stabilizer	λ, 0.1–0.3
Instant chocolate milk stabilization	λ or mixed, 0.1–0.3
Instant milkshake stabilization	λ or mixed, 0.2
Freeze–thaw stable milk gels	ι or mixed, 0.2–0.3
Gelled water desert	κ, ι, 0.5–1
Fruit tart filling/glazing	κ or mixed, 0.7–2
Cooked ham/poultry stabilization	κ, 1–2
Canned meat/fish stabilization	κ, 0.5–1
Low-sugar preserves	Mixed 0.5–1
Instant fruit drink stabilization	λ or mixed 0.2–0.5
Beer clarification	Trace
Salad cream/source stabilization	κ, ι <0.5
Canned/ frozen pet food stabilization	κ, 0.2–0.5

[a]   1% = 10 gL^{-1}.

by itself the latter usually suffers from hardening: this is melting and re-gelling of the toothpaste due to variations of temperature about the melting temperature in warm conditions. The presence of the carrageenan also tends to inhibit enzymic degradation of the paste. κ-carrageenan is also used: it helps to stabilize lower acid preparations.

5.2.7 MICROENCAPSULATION AND IMMOBILIZATION, DRUG DELIVERY AND PHARMACEUTICAL USES

High mechanical strength, chemical inertness and high substrate permeability of κ-carrageenan make it a viable alternative to alginate for microencapsulation of additives. For example, Tosa et al. (1979) looked at the immobilization of several enzyme and cell systems and Table 5.8 illustrates some of their findings. The material to be immobilized was mixed with κ-carrageenan solution prior to gelation, the mixture gelled and the immobilized cells and enzymes assayed as to their activity: activity of the immobilized material was shown to be remarkably high and significantly better than the retention of activity for corresponding material

TABLE 5.8   Activity of Enzymes and Cells after Immobilization in κ-Carrageenan Gel

Enzyme and Microbe	Activity before Immobilization[a]	Activity after Immobilization[a]	Yield (%)
Aminocyclase	20	10	50
Aspartase	650	300	46
Fumarase	360	220	61
Escherichia coli (aspartase)	65,000[b]	30,400	47
Brevibacterium ammoniagenes (fumarase)	9670[c]	5800[d]	60
Streptomyces phaeochromogenes (glucose isomerase)	7910	4280	54

Source: Reprinted from Tosa, T. et al., *Biotechnology Bioengineering*, 21, 1697–1709, 1979.

[a] *Enzyme:* μmol h^{-1} mg^{-1}; *Cell:* μmol h^{-1} g^{-1} (cells).

[b] *After sonication.*

[c] *In the presence of bile extract.*

[d] *After treatment with bile extract.*

immobilized by polyacrylamide gels. As a further example, alcoholic fermentation of glucose by yeast cells immobilized by a range of seaweed polysaccharide gels has been shown by Holcberg and Margalith (1981) to be considerably enhanced compared with the free solution state.

The use of carrageenans in drug delivery has been studied to a great extent and several reviews have been written. It has been shown that matrices based on the carrageenans can increase the drug load, the drug solubility, as well as enabling the release in a good manner also for prolonged period, that is, as a good excipient to for slow release of drugs. This material appears to be of good use, independent of pH and also has an adhesive property being valuable in slow release systems. Carrageenans appear as good materials for use as neutral excipients for drug formulations, as well as use in the food and cosmetics industries (Li et al., 2014; Prajapati et al., 2014; Liu et al., 2015).

5.2.8 BIOLOGICAL EFFECTS

As the carrageenans have wide use in food technology, various studies have been performed to investigate the safety of these compounds. Both McKim (2014) and Weiner (2014) have gone critically through all literature related to the implications of the use of carrageenans for human health and safety and appear to have concluded that no harmful effects have been found so far when taken orally. The majority of the studies have been performed on *in vitro* and animal studies. Campo et al. (2009) presented an overview of the biological properties of the carrageenans, Wang et al. (2012) described the antiviral effects only and Pangestuti and

Kim (2014) also described various biological effects of the carrageenans. Antiviral effects have been studied significantly, and for example an effect against the human papillomavirus (HPV) was shown to be stronger for ι-carrageenan than λ- and κ-carrageenans. The carrageenans were shown to have an anticoagulant activity 1/15th of that of heparin, the cholesterol-lowering effect appears promising, and immunomodulatory effects as well as anti-oxidant effects were seen for these polymers, as well as an anti-tumour effect. The latter may be related to the immunomodulating effects. A hydrolysed part of λ-carrageenan had an increased anti-tumour effect. The conclusions of these reviews are that the carrageenans can be of great importance for industrial and biomedical purposes of various kinds in future.

5.3 AGAR AND AGAROSE

The student may recall agar from school biology days as a culture medium for microbes. It is another red seaweed (*Rhodophyceae*) polysaccharide and is commercially extracted mainly from the *Gelidium*, *Gracilaria* and *Pterociadia* genera. It contains two major components, agarose (also called agaran) which forms strong gels and the more soluble but less commercially interesting non-gelling agaropectin fraction. The latter can be largely removed by a freeze–thaw process. The proportion of agarose in commercial agar varies in general from ~50% up to 90%. The agarose molecule itself is a neutral polysaccharide and not sulphated, although derivatives such as agarose sulphate can be charged. It is very closely related to carrageenan (much more so than alginate) in that the disaccharide repeat is the same, with the exception that the degree of sulphonation is zero (or very small) and the anhydro-α-D-galactose unit is replaced by the L-form:

$$\cdots\rightarrow 3) \ \beta\text{-}D\text{-Gal}p\text{-}(1\rightarrow 4) \ 3, 6\text{-anhydro-}\alpha\text{-}L\text{-Gal}p(1\rightarrow\cdots$$

Figure 5.10 gives the 3D ring structure presentation. The subtle 'mirror-image' difference has been reported from x-ray fibre diffraction studies (Arnott et al., 1974) to give rise to left-handed helical secondary structures as opposed to a right-handed helix in κ- and ι-carrageenan, and the absence of charge allows stronger intrachain interaction phenomena which lead to stronger gels. The helical structure is considered doubled, with one chain half a pitch (i.e. 0.5×1.90 nm) out of phase with the other chain, and has the important feature of a large axial cavity permitting the encapsulation of large amounts of water. On a chemical note, another subtle difference from carrageenan is that up to ~20% of the β-D-Gal*p* residues may contain a methoxy residue at the C6 position.

In common with ι-and κ-carrageenan, during *biosynthesis* the anhydro-form is also missing in the early stages of construction (in the Golgi apparatus). Again, as with carrageenan, kinks in the helical structure – important in the formation of gels – are produced by the occasional replacement of a 3,6-anhydro-α-L–Gal*p* residue by galactose or even galactose-6-sulphate. The less interesting agaropectin fraction is a more complex entity: although the disaccharide repeat $\cdots\rightarrow 3) \ \beta$-D-Gal*p*-(1→4)

Agarose

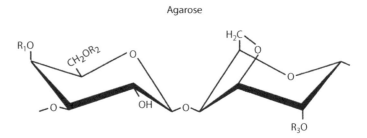

FIGURE 5.10   Agarose disaccharide repeats.

3,6-anhydro-α-L-Gal*p*-(1→··· backbone is essentially the same, the β-D-Gal*p* residues can be replaced to varying degrees by glucuronic acid and in some cases by a pyruvic acid derivative. Also in agaropectin, between 3% and 10% of the 3,6-anhydro-α-L-Gal*p* residues are substituted by sulphated L-Gal*p* residues.

5.3.1 EXTRACTION AND PRODUCTION

Japan – where agar was discovered (reputedly three centuries ago) is, with approximately 2000 tonnes per year, the major supplier of commercial agar. Other major suppliers include Mexico, the United States, Chile, Brazil, Argentina, the western Mediterranean (Spain, Portugal, Morocco), Egypt, India, Korea, the Philippines, New Zealand, Denmark, France and South Africa. Popular species harvested include *Gelidium cartilagineum*, *Gelidium spinulosum* and *Gelidium sesquipedale*. Up to 30% of the dry weight of a plant can be harvested as agar, depending on the season. The yield is highest in summer.

Agar is extracted from the washed and dried seaweed by hot water extraction – agar is largely insoluble at room temperature – adjusted to a pH 5–6, followed by calcium hypochlorite or sodium bisulphite bleaching, followed by rinsing. This is followed by filtration or centrifugation, freezing and subsequent removal of ice, which contains many of the impurities, for purification. Further washing, drying, sterilization, bleaching and a final washing and drying stage can also be used. For applications requiring very high purity further procedures have been devised, involving for example celluloytic-based enzymes, inorganic chemical methods and dialytic desalting.

5.3.2 PROPERTIES

Some fundamental properties are summarized in Table 5.9. The *purity* of commercial samples can be variable: the agarose component is the most commercially useful, and inorganic separation methods can be used to remove the agaropectin fraction, or agaropectin inactivators can be employed, such as aminoalkyldextrans. Other impurities can include

TABLE 5.9 Fundamental Properties: Agarose

Property	Characteristic
Polymer type	Neutral polysaccharide. Alternating repeat
Conformation	Left-handed helical rod. Gross conformation: extra rigid/rigid rod (zone A/B polysaccharide)
Solubility	Insoluble at room temperature (pure agar). Dispersible >90°C
Gelation properties	Extremely potent: It can gel above 0.04%. Melting temperature (at 1.5%) 60°C–100°C. Thermoreversible
Key sites for biomodification	3,6-anhydro-α-L-Galp: anhydro-oxygen bridge (number affects gel strength); C6: degree of sulphonication; C6 (Gal): degree of methylation (affects gelling temperature); 'helical' regions for synergistic interactions with galactomannans; interactions with alginate/starch. (1→4) and (1→3) links are sites for ultrasonic disruption

other saccharides (poly-, oligo- and mono-), starch (which contributes to the turbidity of agar dispersions), dead microbes and trace amounts of heavy metal.

The most notable feature of agar is its gelation properties – agar is probably the most strongest gelling agent known and, depending on the concentration, gel strengths from 500 to over 1000 g cm^{-2} have been reported. These gels are relatively transparent, thermoreversible, stable and strong: although gel strength depends on concentration and temperature, native agar tends to give stronger gels (although less elastic) than purified agarose. Gel strength can be further increased by the addition of starch, alginate or galactomannan, the last of which is thought to interact in a way similar to that for κ- and ι-carrageenans as we have discussed earlier.

5.3.3 POTENTIAL FOR BIOMODIFICATION

The anhydro-oxygen bridge, degree of sulphation and degree of methylation – which affects gelling temperature – are all possible adjustable features, in addition to use of the ordered helical regions for synergistic interactions with galactomannans and with alginate/starch (Guiseley, 1987).

In common with carrageenan, galacto-6-sulphurylases can also be used to convert any 6-O-sulphated groups to the corresponding 3,6-anhydro-α-L-Galp form: this can tend to decrease the number of 'junction zones' and affect gelation properties. Other modifications with potential industrial significance include esterification of primary hydroxyl groups by a variety of organic and inorganic acids, sulphonyl chlorides and isothiocyanates under catalytic conditions.

A possible route of modification is to use enzymes. Those that will catalyze the hydrolysis of agar are classified as α-agarase and β-agarase. The α-agarase will cleave the 1→3 linkages and leave disaccharides related to agarobiose, while the β-agaroses will produce the so-called neoagarooligosaccharides after cleavage of the 1→4 linkages. In both cases, a series of these oligosaccharides will be formed differing with two units from one oligosaccharide to the next in the series. The production of these enzymes have been described by Fu and Kim (2010).

5.3.4 USES OF AGAR AND AGAROSE

Particular use has been made of the mouldability of agar gels: this has been particularly valuable in criminology, plastic surgery and also dentistry although here alginates are now the polysaccharide of preference. In laboratory biochemistry, agarose gel is well known as a separation material for column chromatography or gel filtration, and gel electrophoresis. We focus here on the uses of agar in food, biomedicine and microbiology.

5.3.5 GELLING AND THICKENING IN THE FOOD INDUSTRY

The human alimentary tract by itself is incapable of digesting agar (although some bacteria that are capable of it may be present in the gut). Although its nutritive value is therefore low, it is useful in gelling and emulsifying activities (Table 5.10). An important problem is that agars for food use tend to be in short supply. As a step in addressing this problem, considerable attention has been paid to optimizing agar production and understanding seasonal variations, optimal growth temperatures and so on by studying seaweed growth and agar yields under controlled culture conditions, particularly with regards the *Gracilaria genera* (Bird et al., 1989).

5.3.6 BIOMEDICAL

The high capacity to hydrate – that is to absorb water and the readiness to form flexible mouldable gels – the same property which is so useful in dentistry and criminology, makes agar an ideal non-abrasive medium for use in laxatives. This flexible mouldability is also made use of in surgery as a lubricant. Like some of the other principal marine polysaccharides – notably alginate and also chitosan – its hydratability make it a useful choice as

---

TABLE 5.10   Agar in Food Products

System	Content (%)[a]
Confectionery stabilization and texturization (marshmallows etc.)	~0.05–2
Gelling agent in canned meat products	~0.5–2
Iced lollies and sherbet stabilization and texturization	~0.1
Vegetarian and health food product stabilization	~0.1
Baked goods: reduction in stickiness agent in cookies, icing, meringues etc.	0.1–1

[a] $1\% = 10 \text{ g L}^{-1}$.

either *excipients* or *slow-release forms* in drug delivery, and also in medical physics radiology for use in barium meals.

5.3.7 MICROBIOLOGICAL

Although compared to the other major commercial seaweed polysaccharides its food use is relatively small, one of the main uses of agar – at high purity, free from microbial or microbial interfering substances – is microbiology and indeed has been since Koch's pioneering work at the end of the nineteenth century.

The high gel strength at concentrations between 1% and 2%, thermor-eversibility, uniformity and transparency make high-purity agar gels the classic culture media for microorganisms grown in the laboratory. Agar is generally inert to bacterial metabolism, apart from a few documented exceptions for which other culture support media are recommended. It can also be used in the solution media at lower concentrations (<0.1%) for the growth of anaerobic bacteria.

5.4 FUCOIDANS/FUCOSE-CONTAINING SULPHATED POLYSACCHARIDES

Fucoidan is a terminology used for polysaccharides found in the brown seaweeds, *Phaeophyceae*, being rich in fucose. The fucose containing polysaccharides was first described by Kylin (1913), and was later described in more detail by Conchie and Percival (1950), who isolated the sulphated fucoidan from the brown alga *Fucus vesiculosus*. Other monosaccharides are also normally found in these polymers, both neutral and acidic ones. They are found in the fibrillary cell walls as well as in the intercellular areas in the brown seaweeds.

Their general structure can be described in different ways, but in principle they do consist of a backbone of $1\rightarrow3$ and $1\rightarrow4$ L-fucosyl units. These may be organized as stretches of $1\rightarrow3$ linked units only or as sequences with alternating $1\rightarrow3$ and $1\rightarrow4$ linked units. The L-fucose is in pyranose form; it is often sulphated in positions 2 and 4, and only rarely in position 3. Either single fucosyl units or short oligosaccharide chains may also be attached to the main chain. Depending on the origin of the fucoidan being studied, the detailed structure related to the distribution of the $1\rightarrow3$ and $1\rightarrow4$ linkages, degree of sulphation and the presence of other monosaccharides may vary. Relatively little scientific interest has been reported for the fucoidans until approximately 10 years ago, and since then the importance of these polysaccharides due to their physical and biological properties have increased interest significantly.

5.4.1 EXTRACTION, STRUCTURE AND PRODUCTION

Studies of the fucoidans were not an easy task in the early days. The reason for this was that in addition to the acidic polysaccharide fucoidans, the brown seaweeds were also rich in alginates, consisting only of uronic acids, and thus separation was not easy with the methods available then. Various methods including separation with different cations, like barium and lead salt, led to the first almost pure fucoidan isolated from *Himanthalia lorea* (Percival and Ross, 1950). Newer separation techniques have led to the description of the fucoidans as belonging to three different classes, one group isolated from *Lamnaria* spp., *Analipus* sp., *Caldosiphon* sp. and *Chorda* sp. with a central chain of 1→3 linked alpha-L-fucopyranose units, one group isolated from *Ascophyllum* sp. and *Fucus* sp. with repeating units of a 1→3 and 1→4 linked fucopyranoside units (Patankar et al., 1993; Chevolot et al., 2001) and the third one being very complex as exemplified with the fucoidans isolated from *Turbinaria* sp. having a highly branched structure (Ermakova et al., 2015). The classification of the fucoidans in three groups may also be an oversimplification, as several reviews of the fucoidans have over the years shown that structures vary a lot (Li et al., 2008; Kusaikin et al., 2008; Jiao et al., 2011; Ale et al., 2011; Usonv and Bilan, 2009). In some cases, it has been reported that these fucoidans are so rich in galactose that the term galactofucans should be used (Ermakova et al., 2016; Menshova et al., 2015). Proposed structures can be found in Figure 5.11.

The review by Ale et al. (2011) contains a historic overview of the extraction procedures and chemical analyses of fucoidans from the first ones that were isolated up to the methods used today. In addition, they provide detailed tables showing the differences in structures depending on the isolation processes as well as the origin of the fucoidans. They also conclude that one should characterize the fucoidans as done earlier, that is those with 1→3 linkages as the major ones, those with both 1→3 and 1→4 linkages and the third group being the complex one. All these are sulphated. Thus, those authors have proposed that the term fucoidans should not be used, fucose-containing sulphated polysaccharides (FCSPs) should be the term used for all the different types that can be found.

5.4.2 ENZYMATIC AND CHEMICAL MODIFICATIONS

Due to the complexity of the fucoidans, efforts have been made to find enzymes that could degrade the polymer to give smaller fragments that

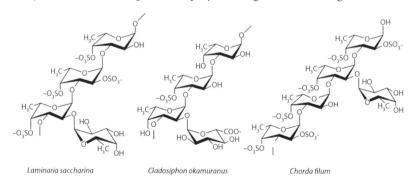

Laminaria saccharina   *Cladosiphon okamuranus*   *Chorda filum*

FIGURE 5.11   Structural motifs for fucans from three different brown algae. (From Ale, M.T. et al., *Marine Drugs*, 9, 2106–2130, 2011.)

are easier to analyse for the real structure. Some of those of great interest have been isolated from marine mollusc and microorganisms. One fucosidase from a mollusc was shown to release terminal L-fucose units, leaving a linear chain containing 1→3 and 1→4 linked units randomly arranged. Another type of enzymes, fucoidanases, from both mollusc and microorganisms, showed the specificity for cleaving unsubstituted 1→3 linked fucose units (Li et al., 2008).

Bacterial fucoidanases have been shown to have a specificity towards highly sulphated 1→3 fucans, while fucoidanases from the marine vertebrates appear to have a preference for degradation of less sulphated 1→3 and 1→4 linked fucans (Kusaykin et al., 2008). Endo-fucoidan-hydrolase from Flavobacteriaceae has been shown to degrade fucoidans, as has other marine bacteria as well. Amongst others a fucoglucuronomannan lyase was isolated from *Fucobacter marina* which gave rise to trisaccharides when subjected to a fucoidan from the brown alga *Kjellmaiella crassifolia*; this fucoidan was then thought to represent a new family of polymers.

Sulphatases have unfortunately not been found to be sufficiently active on this type of polysaccharide apart from one isolated from a mollusc that removed sulphate on position 2 of fucose (Usov and Bilan, 2009).

5.4.3 BIOLOGICAL EFFECTS AND THERAPIES BASED ON FUCOIDANS

Various reviews on the biological effects of fucose-containing polysaccharides as well as structure–activity relations have been published. The effects studied are related to anti-inflammatory, antiviral, antithrombotic and anti-coagulant effects, anti-tumour and immunomodulation, anti-oxidant and also effects against various renal, hepatic and uropathic disorders. These are both *in vitro* and *in vivo* studies. The studies are well described in the reviews written, and thus readers more interested in the various biological effects will find all the details in those. A very interesting review (Fitton, 2011) describes therapies based on the fucoidans. These are based on studies performed *in vivo* as well as *in vitro*. Animal studies showed that the fucoidans, when taken orally, were taken up by the body and when checking the M_w of the material registered in the body of the animals, no degradation had taken place. The *in vivo* studies have in many cases been performed as human clinical trials. In Fitton's review, it is also clearly demonstrated that fucoidans are void of toxicity. The following ailments have been the objects for the studies related to therapeutic use of the fucoidans: inflammation and fibrosis, osteoarthritis, surgical adhesion, liver fibrosis, radiation, stem cells, blood homeostasis, neuronal protection, viral, bacterial and parasite infections and renal diseases.

It is also interesting that the effects vary with the source of the polysaccharide, showing that details in the structure are important for the effects observed. The last review is from 2013, thus newer publications dealing with biological activities are also referenced here (Senthilkumar et al., 2013; Ermakova et al., 2015; Menshova et al., 2015; Ermakova et al., 2015; Trincone et al., 2015; Hifney et al., 2016).

5.5 CHITOSAN AND CHITIN DERIVATIVES

Chitosans – the most important solubilized derivative of chitin (Old Greek: *chiton*: tunic or envelope) – are receiving ever-increasing interest. This is principally because of their polycationic properties, biodegradability and low antigenicity.

The basic structure of chitin is very similar to that of cellulose, also insoluble in aqueous solvents: cellulose has $\beta(1\rightarrow4)$-linked D-glucopyranosyl residues whereas chitin has $\beta(1\rightarrow4)$-linked N-acetyl-D-glucosamine residues:

$$...\rightarrow4)\ \beta\text{-}D\text{-}N\text{-acetyl-Glc}p\ (1\rightarrow...$$

In chitosan, the N-acetyl group is replaced by NH_2 to varying degrees. Except at high pH, this latter group exists as NH_3^+ and it is the range of properties that can be obtained by controlling the degree of acetylation, F_A (=0, fully deacetylated and =1, fully acetylated, i.e. 'chitin') together with its low antigenicity and relative abundance that makes this one of the most biotechnologically exciting polysaccharides. Figure 5.12 shows the Haworth projection of an acetylated residue coupled to a deacetylated residue. Although chitosan exists as a single chain, because of the charge repulsion effects it adopts a rigid linear structure in solution, that is, it is a zone B polysaccharide approaching semi-flexible (zone C) at lower charge density (higher F_A). For further details on the action of chitin deacetylases and other aspects of chitin enzymology, the interested student should consult two well-established volumes by Muzzarelli (1993, 1996).

5.5.1 EXTRACTION AND PRODUCTION

A whole variety of possible sources exist from two main classes of organisms: crustaceans and mushrooms. The main source is the shells of shrimp, crab and lobster and is an excellent example of use of waste materials, in this case from the seafood industry: unused shells and heads are rich in chitin (15%–30%, the rest mostly calcium carbonate). The other popular source is the mycelli (containing up to 20% chitin) from mushrooms and other fungi. About an average of 30%–60% of the cuticles of insects is chitin. The current world market is approximately

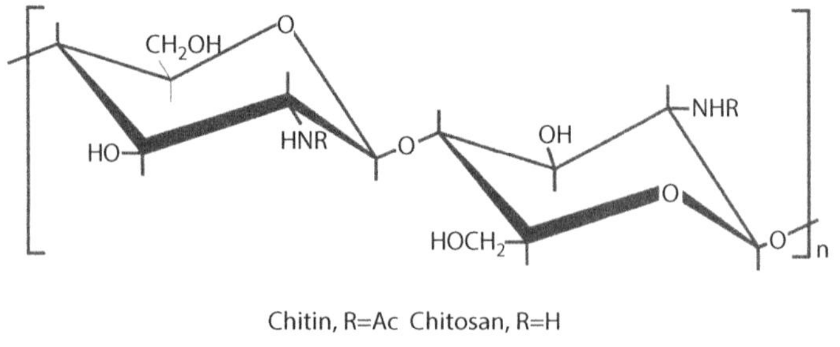

Chitin, R=Ac Chitosan, R=H

FIGURE 5.12   Chitosan structural units.

2000 tonnes per anum. Japan – where chitosan is food grade approved – is the biggest user (700,000 kg p.a.) followed by Europe (250,000 kg p.a.) and the United States (100,000 kg p.a.) where it is not (yet) food grade approved.

To produce chitosan commercially from marine chitin, shells are ground and filtered followed by a 24-hour demineralization by HCl in a countercurrent exchanger. Proteins are then removed by the usual protease treatment (pepsin or trypsin digestion) or using 5% NaOH at 90°C for about 30 minutes. Peroxide depigmentation is then applied, then comes the key step of controlled deacetylation using alkali treatment at the higher temperatures used for deproteinization, using potassium hydroxide, ethanol and ethylene glycol. Deacetylation is conventionally taken to 80%–85% (i.e. to an F_A of 0.2), but usually no further because the alkali treatment also leads to depolymerization as monitored by a drop in viscosity (i.e. molecular weight), and commercial manufacturers usually look for a viscosity of 1200 cP for a concentration of 12.5 gL^{-1} (1.25%). In Europe, a commercial manufacturer is Novamatrix (Sandvika, Norway).

5.5.2 PROPERTIES

Table 5.11 summarizes the fundamental properties of a popular commercial chitosan of relatively high degree of deacetylation (F_A ~0.11), although as with other polysaccharides it has to be remembered there can be considerable batch variation in properties. Again, although the weight average M_w can be readily obtained using conventional sedimentation-diffusion, sedimentation equilibrium or light scattering types of analysis extraction of molecular weight distributions has proved difficult, because separation columns do not function well with polycationic materials.

TABLE 5.11   Fundamental Properties of a Commercial Chitosan: Sea-Cure + 210 (Pronova Ltd., Norway)

Property	Characteristic
Polymer type	Polycationic; interrupted repeat
Conformation type	Zone C (semi-flexible coil)/ Zone B (rod)
Degree of acetylation F_A	0.11
Solubility	pH < 6
Molecular weight	~(160,000 ± 10,000) Da
Intrinsic viscosity [η]	~(540 ± 20) mL g^{-1}
Sedimentation coefficient, $s_{20,w}$	(1.41 ± 0.05) S
Key sites for biomodification	N-Acetyl group (alteration of F_A value); C6 OH group (etherification)

Source: Data from Errington, N. et al., *International Journal of Biological Macromolecules,* 15, 113–117, 1993.

5.5.3 BIOMODIFICATION

The key site for biomodification is the *N*-acetyl group, that is control of the degree of acetylation. The method of Hirano and Yamaguchi (1976) achieves this by reacetylating to varying degrees 100% deacetylated ($F_A = 0$) chitosan. It is also the site for enzyme degradation by lysozyme or other chitinase/chitosanase activity to give chitosans of smaller chain lengths and, ultimately, oligosaccharides: this is particularly important for biomedical applications. Other methods for reducing the polymer chain length include fluorohydrolysis (using HF) and sonication. Yabuki and coworkers (Yabuki et al., 1988) have shown a strong dependence of enzyme susceptibility on the F_A value, with 0.0–0.5 susceptible and a maximum at ~0.2.

The other popular site for substitution is the hydroxyl group on the C6 atom and this is the site for esterification or, better, etherification to give one of the most popular of chitin derivatives, 6-*O*-carboxymethylchitin ($F_A = 1$) (Nishimura et al., 1986) and the related 6-*O*-carboxymethylchitosan ($F_A < 1$). Figure 5.13 shows the potential use as a carrier for the controlled

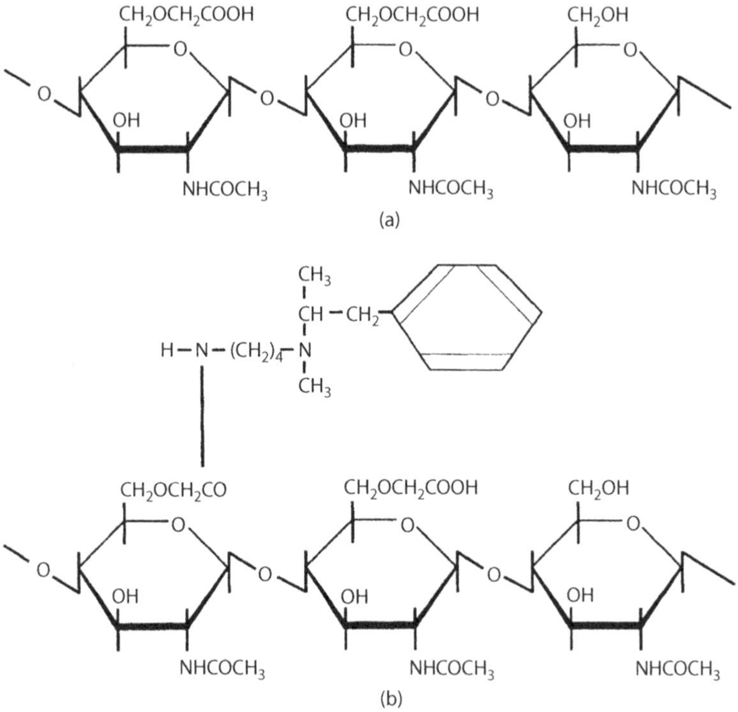

FIGURE 5.13 (a) 6-*O*-carboxy-methylchitin. (b) The acidic residue can itself be a site for drug attachment and the first residue is shown linked via a spacer hydrocarbon to a hapten molecule. (Reprinted and adapted from Skjåk-Bræk, G. et al., *Chitin and Chitosan*, Baba, S. et al., Controlled release and hydrolysis prodrug using carboxymethyl-chitin as a drug carrier, pp. 703–711, 1989, with permission from Elsevier.)

release and hydrolysis of a drug using the acidic residues. This potentially offers a great advantage over unsubstituted chitosans in that it is soluble at normal pHs. Attempts have been made to produce water-soluble (as opposed to acid-soluble) chitosans: in this case, the lactic acid salt of chitosan seems to have met with some success.

5.5.4 USES OF CHITOSAN

5.5.4.1 FOOD BIOTECHNOLOGY: FINING AGENTS FOR FRUIT JUICES AND BEERS

Particulates in fruit juices are often negatively charged, deriving largely from pectins. The usual clarification procedure is via sedimentation tanks or centrifugation. The speed of clarification can be greatly enhanced by using positively charged chitosans as fining agents, yielding larger flocculates which sediment faster. Chitosans' dye binding properties can also be put to use for the recovery of beta-carotene in carrot juice processing.

5.5.4.2 GEL-MATRIX TECHNOLOGY (MICROENCAPSULATION)

Chitosan-based gel beads have been applied for the entrapment of microbes, plant cells and mammalian cells as well as enzymes. Of particular interest are combined chitosan/alginate or chitosan/carrageenan and chitosan/carboxymethylcellulose systems which appear to provide greater controllability in the entrapment and release mechanisms. Figure 5.14 shows the sustained release properties of a 793-Da dye from an alginate–chitosan capsule.

5.5.4.3 POROUS BEAD TECHNOLOGY: CHELATION AND AFFINITY CHROMATOGRAPHY

Chitosan is known to interact with heavy metals via its amino group. This has led to the rationale for its use in porous bead technology for the chelation of heavy metal ions and in the design of affinity columns for immobilizing proteins. For example, Seo and Kinemura (1989) observed the immobilization of 40-mg bovine serum albumin and 46-mg haemoglobin from passage in each case of 1 mL of 10 mg mL^{-1} protein solution through 'Chitopearl beads' (~10 µm in diameter) developed from an $F_A = 0.05$–0.2 chitosan.

5.5.4.4 PATHOGEN RESISTANCE IN CROPS

Chitosan, from the cell walls of some fungi, is known to play an important role in the relation of the fungus with its host plant. Diffusion of chitosans into the host plant tissue can induce a host response – including disease-resistant host genes against the fungus: application of external chitosan to crops can enhance their resistances to fungal attack. These mechanisms are being studied and appear to involve complexes between the host DNA and the chitosan, probably due to anion–cation interaction.

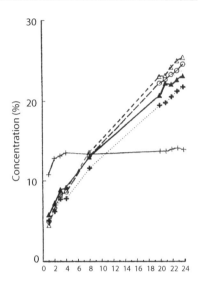

FIGURE 5.14   Sustained release of a 793-Da dye from alginate–chitosan capsules as a function of chitosan (chitosan lactate) concentration. Crosses: 0 mg mL^{-1}; open triangles: 2.5 mg mL^{-1}; circles: 5 mg mL^{-1}; closed triangles: 7.5 mg mL^{-1}; thick crosses: 10 mg mL^{-1}. (Reprinted from Skjåk-Bræk, G. et al., *Chitin and Chitosan*, Knorr, D. et al., Potential of acid soluble and water soluble chitosans in biotechnology, pp. 101–118, 1989, with permission from Elsevier.)

5.5.4.5 GENE THERAPY

DNA–chitosan interaction is also being explored for use in gene therapy – insertion of DNA into patients' cells by using chitosan to condense DNA (by analogy with histone action) into micro-sizes. Chitosan appears to be an attractive alternative to other cationic condensing agents (such as spermidine) which can be considerably toxic.

5.5.4.6 HEMOSTATIS AND WOUND HEALING TECHNOLOGY

The ability of chitosan to bind to cells has been known for a long time (Muzzarelli et al., 1989). Besides being useful for spermicidal and anti-cancer technology, the red-cell binding activity – which acts independently of the body's own blood clotting cascade – has been extensively explored, although reports on the useful effects for wound healing applications are less conclusive compared with alginate. Carboxymethylchitin has also been strongly considered.

5.5.4.7 BANDAGE CONTACT LENSES (EYE THERAPY)

The possible wound healing and antimicrobial effects of chitosan together with its excellent film forming properties have made it attractive for 'bandage contact lenses', particularly for possible improvements in corneal healing.

5.5.4.8 FILM-FORMING AGENTS IN COSMETICS

Film formation is taken advantage of by the cosmetic industry for hair-setting sprays and creams, nail varnishes and skin creams. The additional thickening properties of chitosan also make it attractive for use in shampoos and hair conditioners.

5.5.4.9 MUCOADHESIVES

The oral route is a popular method of administering drugs (Fiebrig et al., 1995). This requires passage through the alimentary tract and eventual absorption through the mucosal membrane, usually in the proximal small intestine. Unfortunately, the amount of drug actually delivered, the 'bioavailability', can be very much smaller than that ingested because of (1) too rapid a transit of the drug-containing system past the ideal absorption site, (2) rapid degradation of the drug in the gastrointestinal tract once it has been released and (3) low transmucosal permeability due to the size, ionization, solubility or other characteristics of the drug molecule. Chitosans provide an attractive mucoadhesive, molecular brake or delay for a drug to be encapsulated because of their polycationic complementarity to the polyanionic mucins that constitute the key macromolecular component of mucus (Harding, 2003, 2006). Laboratory trials appear to have proven the success of chitosan as a mucoadhesive using a range of macromolecular characterization methods including co-sedimentation analysis and electron microscopy. Figure 5.15a shows an electron micrograph of a mucin–chitosan complex, and Figure 5.15b shows the corresponding image with the chitosan component conjugated with colloidal gold for identification purposes: the latter shows the chitosan clearly in the interior of the large complex. Figure 1.15 in Chapter 1 gives the corresponding images from atomic force microscopy.

   Although the strong mucoadhesive ability of chitosan is well established, its ability to dock and undock a drug from the chitosan formulation is still being explored. Chitosans appear to have the interesting 'bonus' effect of actually enhancing the ability of the mucosal membrane to absorb a drag molecule as large as insulin: this has been demonstrated for nasal epithelia, and a mechanism has been suggested where the chitosan reduces the beat frequency of the cilia assisting a transient opening of pericellular junctions.

5.5.4.10 OTHER MEDICINAL USES

The chitosans are well known as products in wound healing, an important effect is the accelerating effect bandages based on chitosan has been shown to have. Artificial kidney membranes are another use of this material, and as is the case for many other marine polysaccharides, the chitosans have also gained an increased interest in drug delivery, especially in slow release systems. Chitosan has both haemostatic and anti-coagulant effect, properties of importance in wound healing, and has also been tried as artificial skin.

   Recently, chitooligosaccharides (COS) have been produced, and their use in the medical field is growing. This is basically due to their solubility

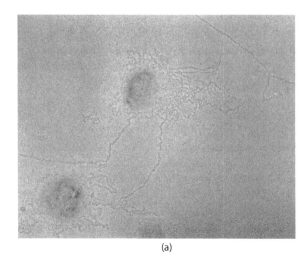

(a)

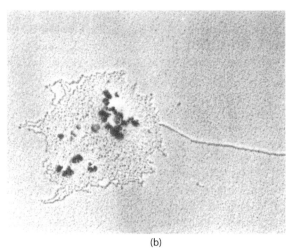

(b)

FIGURE 5.15   Electron microscopy (magnification ×60,000) of a chitosan–mucoadhesive system. Chitosan: Sea Cure + 210 (F_A = 0.11). Mucin: pig gastric mucin. (a) Electron micrograph of pig gastric mucin complexed with chitosan. (b) Electron micrograph of pig gastric mucin complexed with chitosan conjugated with colloidal gold (appearing as the very dark areas) shows that the chitosan is chiefly in the interior of the complex. As a further gauge of the magnitude of the mucoadhesive effect, the sedimentation coefficient of the chitosan and mucin is ~2 and 50 S, respectively: that of the complex is ~2000 S. (Reprinted from *Carbohydrate Polymers*, 33, Fiebrig, I. et al., Colloidal gold and colloidal gold labelled wheat germ agglutinin as molecular probes for identification in mucin/chitosan complexes, 91–99, 1997, with permission from Elsevier.)

which makes them easier to handle than the native polymer chitin and the deacetylated chitosan. Derivatives have been made to produce interesting bioactive substances, like the gallyl-oligosaccharide, which has anti-oxidative effects. The chitins, chitosans and derivatives do regulate immune responses, and due to this, the compounds have also found use as adjuvants in cancer vaccines.

FURTHER READING

Ahmed, A.B.A., Adel, M., Karimi, P. and Peidayesh, M. (2014) Pharmaceutical, cosmeceutical and traditional applications of marine carbohydrates. *Advances in Food and Nutrition Research*, 73, 197–220.

Almutairi, F.M., Adams, G.G., Kök, M.S., Lawson, C.J., Gahler, R., Wood, S., Foster, T.J., Rowe, A.J. and Harding, S.E. (2013) An analytical ultracentrifugation based study on the conformation of lambda carrageenan in aqueous solution, *Carbohydrate Polymers*, 97, 203–209.

Cardoso, M.J., Costa, R.R. and Mano, J.F. (2016) Marine origin polysaccharides in drug delivery systems, *Marine Drugs*, 14, 34.

D'Ayala, G.G., Malinconico, M. and Laureienzo, P. (2008) Marine derived polysaccharides for biomedical applications: Chemical modification approaches, *Molecules*, 13, 2069–2106.

Dibenedetto, A. (2011) The potential of aquatic biomass for CO_2 – enhanced fixation and energy production, *Greenhouse Gases Science and Technology*, 1, 58–71.

Lapasin, R. and Pricl, S. (1995) *Rheology of Industrial Polysaccharides. Theory and Applications*, Chaps. 1 and 2, London: Blackie.

Laurienzo, P. (2010) Marine polysaccharides in pharmaceutical applications: An overview. *Marine Drugs*, 8, 2435–2465.

Makkar, H.P.S., Tran, G., Heuze, V., Giger-Reverdin, S., Lessire, M. and Lebas, F. (2016) Seaweeds for livestock diets: A review, *Animal Feed Science and Technology*, 212, 1–17.

Rhein-Knudsen, N., Ale, M.T. and Meyer, A.S. (2015) Seaweed hydrocolloid production: An update on enzyme assisted extraction and modification technologies, *Marine Drugs*, 13, 3340–3359.

Rinaudo, M. (2006) Characterisation and properties of some polysaccharides used as biomaterial, *Macromolecular Symposia*, 245–246, 549–557.

Sudha, P.N., Nithya, A.R. and Vijayalakshmi, K. (2014) Industrial application of marine carbohydrates, *Advances in Food and Nutrition Research*, 73, 145–181.

Alginates

Clark, K. (1993) Algin, in Whistler, R.L. and BeMiller, J.N. (eds) *Industrial Gums*, Chap. 6, New York, NY: Academic Press.

Lee, K.Y. and Mooney, D.J. (2012) Alginate: Properties and biomedical applications, *Progress in Polymer Science*, 37, 106–126.

O'Loughlin, J.A., Bruder, J.M. and Lysaght, M.J. (2004) Oral administration of biochemically active microcapsules to treat uremia: New insights into an old approach, *Journal of Biomaterials Science, Polymer Edition*, 15, 1447–1461.

Pawar, S.N. and Edgar, K.J. (2012) Alginate derivatization: A review of chemistry, properties and applications, *Biomaterials*, 33, 3279–3305.

Skjåk-Bræk, G., Flo, T., Halaas, Ø. and Espevik, T. (2000) Immune stimulating properties of di-equatorial β(1–4) linked poly-uronides, *Bioactive Carbohydrate Polymers*, 44, 85–93.

Various authors. (1996) in *Carbohydrates in Europe*, Vol. 14, May 1996.

Yang, J.-S., Xie, Y.-J. and He, W. (2012) Research progress on chemical modifications of alginate: A review, *Carbohydrate Polymers*, 84, 33–39.

Carrageenans

Campo, V.L., Kawano, D.F., Da Silva, Jr., D.B. and Carvalho, I. (2009) Carrageenans: biological properties, chemical modifications and structure analysis—A review, *Carbohydrate Polymers*, 77, 167–180.

Li, L., Ni, R., Shao, Y. and Mao, S. (2014) Carrageenan and its application in drug delivery, *Carbohydrate Polymers*, 103, 1–11.

Liu, J., Zhan, X., Wan, J., Wang, Y. and Wand, C. (2015) Review for carrageenan-based pharmaceutical biomaterials: Favourable physical features versus adverse biological effects, *Carbohydrate Polymers*, 121, 27–36.

McKim, J.M. (2014) Food additive carrageenan: Part I: A critical review of carrageenan in vitro studies, potential pitfalls, and implications for human health and safety, *Critical Reviews in Toxicology*, 44, 211–243.

Pangestuti, R. and Kim, S.-K. (2014) Biological activities of carrageenan, *Advances in Food and Nutrition Research*, 72, 113–124.

Prajapati, V.D., Maheriya, P.M., Jani, G.K. and Solanki, H.K. (2014) Carrageenan: A natural seaweed polysaccharide and its applications, *Carbohydrate Polymer*, 105, 97–112.

Therkelsen, G.H. (1993) Carrageenan, in Whistler, R.L. and BeMiller, J.N. (eds) *Industrial Gums*, Chap. 7, New York, NY: Academic Press.

Wang, W., Wang, S.-X and Guan, H.-S. (2012) The antiviral activities and mechanisms of marine polysaccharides: An overview, *Marine Drugs*, 10, 2795–2816.

Weiner, M.L. (2014) Food additive carrageenan: Part II: A critical review of carrageenan in vivo safety studies, *Critical Reviews in Toxicology*, 44, 244–269.

Agar

Fu, X.T. and Kim, S.M. (2010) Agarase: Review of major sources, categories, purification methods, enzyme characteristics and applications, *Marine Drugs*, 8, 200–218.

Selby, H.H. and Whistler, R.L. (1993) Agar, in Whistler, R.L. and BeMiller, J.N. (eds) *Industrial Gums*, Chap. 7, New York, NY: Academic Press.

Fucoidans

Ale, M.T., Mikkelsen, J.D. and Meyer, A.S. (2011) Important determinants for fucoidan bioactivity: A critical review of structure-function relations and extraction methods for fucose-containing sulfated polysaccharides from brown seaweeds, *Marine Drugs*, 9, 2106–2130.

Chevolot, L., Mulloy, B., Ratiscol, J., Foucault, A. and Colliec-Jouault, S. (2001) A disaccharide repeat unit is the major structure in fucoidans from two species of brown algae, *Carbohydrate Research*, 330, 529–535.

Conchie, J. and Percival, E.G.V. (1950) Fucoidin. Part II. The hydrolysis of a methylated fucoidin prepared from *Fucus vesiculosus*, *Journal of Chemical Society*, 827–832.

Ermakova, S., Kusaykin, M., Trincone, A. and Zvyagintseva, T. (2015) Are multifunctional marine polysaccharides a myth or reality? *Frontiers in Chemistry*, 3, 39.

Ermakova, S.P., Menshova, R.V., Anastyuk, S.D., Malyarenko, O.S., Zakharenko, A.M., Thinh, P.D., Ly, B.M. and Zvyagintseva, T.N. (2016). Structure, chemical and enzymatic modification, and anticancer activity of polysaccharides from brown alga *Turbinaria ornate*, *Journal of Applied Phycology*, 28(4), 2495–2505.

Fitton, J.H. (2011) Therapies from fucoidans; multifunctional polymers, *Marine Drugs*, 9, 1731–1760.

Hifney, A.F., Fawzy, M.A., Abdel-Gawad, K.M. and Gomaa, M. (2016) Industrial optimization of fucoidan extraction from *Sargassum* sp. and its potential antioxidant and emulsifying activities, *Food Hydrocolloids*, 54, 77–88.

Jiao, G., Yu, G., Zhang, J. and Ewart, S. (2011) Chemical structures and bioactivities of sulfated polysaccharides from marine algae, *Marine Drugs*, 9, 196–223.

Kusaykin, M., Bakunina, I., Sova, V., Ermakova, S., Kuznetsova, T., Besednova, N., Zaporozhets, T. and Zvyagitseva, T. (2008) Structure, biological activity, and enzymatic transformation of fucoidans from brown seaweeds, *Journal of Biotechnology*, 3, 904–915.

Kylin, H. (1913) Biochemistry of sea algae, *Physical Chemistry*, 83, 171–197.

Li, B., Lu, F., Wei, X. and Zhao, R. (2008) Fucoidan: Structure and bioactivity, *Molecules*, 13, 1671–1695.

Menshova, R.V., Anastyuk, S.D., Ermakova, S.P., Shevencho, N.M., Isakov, V.I. and Zvyagitseva, T.N. (2015) Structure and anticancer activity in vitro of sulfated galactofucan from brown alga Alaria angusta, *Carbohydrate Polymers*, 132, 118–125.

Patankar, M.S., Oehninger, S., Barnett, T., Williams, R.L. and Clark, G.F. (1993) A revised structure for fucoidan may explain some of its biological activities, *Journal of Biological Chemistry*, 268, 21770–21776.

Percival, E.G.V. and Ross, A.G. (1950) Fucoidin. Part 1. The isolation and purification of fucoidin from brown seaweeds, *Journal of Chemical Society*, 717–720.

Senthilkumar, K., Manivasagan, P. and Venkatesa, N.J. (2013) Brown seaweed fucoidan: Biological activity and apoptosis, growth signalling mechanism in cancer, *International Journal of Biological Macromolecules*, 60, 366–374.

Trincone, A., Kusyakin, M. and Ermakova, S. (2015) Editorial: Marine biomolecules, *Frontiers in Chemistry*, 3, 52.

Usov, A.I. and Bilan, M.I. (2009) Fucoidans—Sulfated polysaccharides from brown algae, *Russian Chemical Reviews*, 78, 785–799.

Chitosan and Chitin derivatives

Azuma, K., Izumi, R., Osaki, T., Ifuku, S., Morimoto, M., Saimoto, H., Minami, S and Okamoto, Y. (2015) Chitin, chitosan and its derivatives for wound healing: Old and new material, *Journal of Functional Biomaterials*, 6, 104–142.

Jayakumar, R., Prabaharan, M., Sudheesh Kumar, P.T., Nair, S.V., Furuike, T. and Tamura, H. (2011) Novel chitin and chitosan materials in wound dressing, in Laskovski, N. (ed) *Biomedical Engineering, Trends in Materials Science*, pp. 3–24.

Li, X., Min, M., Du, N., Gu, Y., Hode, T., Naylor, M., Chen, D., Nordquist, R.E. and Chen, W.R. (2013) Chitin, chitosans and glycated chitosan regulate immune responses: The novel adjuvants for cancer vaccine, *Clinical and Developmental Immunology*, 2013, article ID 387023.

Lodhi, G., Kim, Y.-S., Hwang, J.-W., Kim, S.-H., Jeon, Y.-J., Ahn, C.-B., Moon, S.-H., Jeon, B.-T. and Park, P.-J. (2014) Chitooligosaccharides and its derivatives: Preparation and biological applications, *Biomed Research International*, 2014, article ID 654913.

Roberts, G.A.F. (1991) *Chitin Chemistry*, London: Macmillan Press.

Sahoo, D. and Nayak, P.L. (2011) Chitosan, the most valuable derivative of chitin, in Kalia, S. and Averous, L. (eds) *Biopolymers: Biomedical and Environmental Application*, pp. 129–166.

Skjåk-Bræk, G., Anthonsen, T. and Sandford, P. (eds) (1989) *Chitin and Chitosan*, Amsterdam, the Netherlands: Elsevier.

Whistler, R.L. (1993) Chitin, in Whistler, R.L. and BeMiller, J.N. (eds) *Industrial Gums*, Chap. 22, New York, NY: Academic Press.

SPECIFIC REFERENCES

Almutairi, F.M., Adams, G.G., Kök, M.S., Lawson, C.J., Gahler, R., Wood, S., Foster, T.J., Rowe, A.J. and Harding, S.E. (2013) An analytical ultracentrifugation based study on the conformation of lambda carrageenan in aqueous solution. *Carbohydrate Polymers* 97, 203–209.

Arnott, S., Fuller, A., Scott, W.E., Dea, I.C.M., Moorhouse, R. and Rees, D.A. (1974) The agarose double helix and its function in agarose gel structure, *Journal of Molecular Biology*, 90, 269–284.

Baba, S., Uraki, Y., Miura, Y. and Tokura, S. (1989) Controlled release and hydrolysis prodrug using carboxymethyl-chitin as a drug carrier, in Skjåk-Bræk, G., Anthonsen, T. and Sandford, P. (eds) *Chitin and Chitosan*, pp. 703–711, Amsterdam, the Netherlands: Elsevier.

Bird, K., Pendoley, K. and Koehn, F. (1989) Variability in agar gel behaviour and chemistry as affected by algal growth under different environmental conditions, in Cresenzi, V., Dea, I.C.M., Paoletti, S., Stivala, S.S. and Sutherland, I.W. (eds) *Biomedical and Biotechnological Advances in Industrial Polysaccharides*, pp. 365–73, New York, NY: Gordon and Breach.

Craigie, J.S. and Wong, K.F. (1979) Carrageenan biosynthesis, *Proceeding International Seaweed Symposium*, (Santa Barbara, USA) vol. 9, pp. 369–311

Errington, N., Harding, S.E., Vårum, K.M. and Illum, L. (1993) Hydrodynamic characterization of chitosans varying in degree of acetylation, *International Journal of Biological Macromolecules*, 15, 113–117.

Ertesvag, H., Valla, S. and Skjåk-Bræk, G. (1996) Genetics and synthesis of alginates, *Carbohydrates in Europe*, 14, 14–18.

Fiebrig, I., Davis, S.S. and Harding, S.E. (1995) Methods used to develop mucoadhesive drug delivery systems: Bioadhesion in the gasterointestinal tract, in Harding, S.E., Hill, S.E. and Mitchell, J.R. (eds) *Biopolymer Mixtures*, Chap. 18, Nottingham: Nottingham University Press.

Fiebrig, I., Vårum, K.M., Harding, S.E., Davis, S.S. and Stokke, B.T. (1997) Colloidal gold and colloidal gold labelled wheat germ agglutinin as molecular probes for identification in mucin/chitosan complexes, *Carbohydrate Polymers*, 33, 91–99.

Fischer, F.G. and Dörfel, H. (1955) Die polyuronsäuren der braunlagen, *Zeitschrift fur Physiologische Chemie*, 302, 186–205.

Grant, G.T., Morris, E.R., Rees, D.A., Smith, P.J.C. and Thom, D. (1973) Biological interactions between polysaccharides and divalent cations: The egg-box model, *FEBS Letters*, 32, 195–198.

Guiseley, K. B. (1987) Natural and synthetic derivatives of agarose and their use in biochemical separation, in Yalpani, M. (ed) *Industrial Polysaccharides: Genetic Engineering, Structure/Property Relations and Applications*, pp. 139–48, Amsterdam, the Netherlands: Elsevier.

Harding, S.E. (2003). Mucoadhesive interactions, *Biochemical Society Transactions*, 31, 1036–1041.

Harding, S.E. (2006). Trends in mucoadhesive analysis, *Trends in Food Science and Technology*, 17, 255–262.

Harding, S.E., Day, K., Dhami, R. and Lowe, P.M. (1997) Further observations on the size, shape and hydration of kappa-carrageenan in dilute solution, *Carbohydrate Polymers*, 32,81–87.

Hirano, S. and Yamaguchi, R. (1976) *N*-Acetylchitosan gel: A polyhydrate of chitin, *Biopolymers*, 15, 1685–1691.

Holcberg, I.B. and Margalith, P. (1981) Alcohol fermentation by immobilized yeast at high sugar concentrations, *European Journal of Applied Microbiology Biotechnology*, 13, 133–140.

Horton, J.C., Harding, S.E. and Mitchell, J.R. (1991a) Gel permeation chromatography multi-angle laser light scattering characterization of the molecular mass distribution of 'Pronova' sodium alginate, *Biochemical Society Transaction*, 19, 510–511.

Horton, J.C., Harding, S.E., Mitchell, J.R. and Morton-Holmes, D.F. (1991b) Thermodynamic non-ideality of dilute solutions of sodium alginate studied by sedimentation equilibrium, *Food Hydrocolloids*, 5, 125–127.

Knorr, D., Beaumont, M.D. and Pandya, Y. (1989) Potential of acid soluble and water soluble chitosans in biotechnology, in Skjäk-Braek, G., Anthonsen, T. and Sandford, P. (eds) *Chitin and Chitosan*, pp. 101–118, Amsterdam, the Netherlands: Elsevier.

Lama, D. and Segre, A.L. (1980) Structural analysis of neocarrabiose by X-ray crystallography, NMR spectroscopy and molecular mechanics calculations, in Crescenzi, V., Dea, I.C.M., Paoletti, S., Stivala, S.S. and Sutherland, I.W. (eds) *Biomedical and Biotechnological Advances in Industrial Polysaccharides*, pp. 459–467, New York, NY: Gordon & Breach.

Muzzarelli, R.A.A. (ed) (1993) *Chitin Enzymology*, Vol. 1, Atec, Grottammare, Italy.

Muzzarelli, R.A.A. (ed) (1996) *Chitin Enzymology*, Vol. 2, Atec, Grottammare, Italy.

Muzzarelli, R.A.A., Biagini, G., Damadei, A., Pugnaloni, A. and Da Lio, J. (1989) Chitosans and other polysaccharides as wound dressing materials, in Crescenzi, V., Dea, I.C.M., Paoletti, S., Stivala, S.S. and Sutherland, I.W. (eds) *Biomedical and Biotechnological Advances in Industrial Polysaccharides*, pp. 77–88, New York, NY: Gordon & Breach.

Nishimura, S., Nishi, N., Tokura, S., Nishimura, K. and Azuma, I. (1986) Bioactive chitin derivatives. Activation of mouse-macrophages by *O*-(carboxymethyl) chitins, *Carbohydrate Research*, 146, 251–258.

Onsoyen, E. (1996) Commercial applications of alginates, *Carbohydrates in Europe*, 14, 26–31.

Padol, A.M. Maurstad, G., Draget, K.I., and Stokke, B.T. (2015) Delaying cluster growth of ionotropic induced alginate gelation by oligoguluronate. *Carbohydrate Polymers*, 133, 126–134.

Richardson, T., Kumta, P.N. and Banerjee I. (2014) Alginate encapsulation of human embryonic stem cells to enhance directed differentiation to pancreatic islet-like cells. *Tissue Engineering Part A.*, 3198–3211.

Schacht, E., Vandichel, J.C., Lemahieu, A., De Rooze, N. and Vansteenkiste, S. (1993) The use of gelatin and alginate for the immobilization of bioactive agents, in Karsa, D.R. and Stephenson, R.A. (eds) *Encapsulation and Controlled Release*, pp. 18–34, Cambridge: Royal Society of Chemistry.

Schmidt, R.J. (1991) Alginates in wound management, in Smidsrød, O., Skjåk-Bræk, G. and Draget, K.I. (eds) *First Advanced Course on Alginates and Their Applications*, Norway: University of Trondheim.

Seo, H. and Kinemura, Y. (1989) Preparation and some properties of chitosan porous beads, in Skjåk-Bræk, G., Anthonsen, T. and Sandford, P. (eds) *Chitin and Chitosan*, pp. 585–603, Amsterdam, the Netherlands: Elsevier.

Skaugrud, O. (1993) Drug delivery systems with alginate and chitosan, in Karsa, D.R. and Stephenson, R.A. (eds) *Excipients and Delivery Systems for Pharmaceutical Applications*, pp. 96–107, Cambridge: Royal Society of Chemistry.

Skjåk-Bræk, G. and Espevik, T. (1996) Application of alginate gels in biotechnology and biomedicine, *Carbohydrates in Europe*, 14, 19–25.

Skjåk-Bræk, G. and Martinsen, A. (1991) Biotechnological application of some marine polysaccharides, in Blinden, G. and Guiery, M. (eds) *Seaweed Resources in Europe—Use and Potential*, pp. 221–57, London: John Wiley & Sons.

Skjåk-Bræk, G., Smidsrød, O., Eklund, T., Huseby, K.O. and Kvam, B. (1996) Chemical modifications of alginates: C-5 epimerisation of polymer linked d-mannuronic acid by treatment with supercritical carbon dioxide. US Patent 4,990,601.

Slootmaekers, D., van Dijk, J.A.P.P., Varkevisser, F.A., Bloys van Treslong, C.J. and Reynaers, H. (1991) Molecular characterisation of κ- and ι- carrageenan by gel permeation chromatography, light scattering, sedimentation analysis and osmometry, *Biophysical Chemistry*, 41, 51–59.

Thu, B., Espevik, P., Soon-Shiong, P., Smidsrød, O. and Skjåk-Bræk, G. (1996a) Alginate poly-lysine capsules II. Functional properties, *Biomaterials*, 17, 1069–1079.

Thu, B., Smidsrød, O. and Skjåk-Bræk, G. (1996b) Alginate gels. Some structure–function correlations relevant to their use as immobilization matrix for cells, in Wijffels, R.J., Buitelaar, R.M., Bucke, C. and Tramper, J. (eds) *Immobilized Cells: Basics and Applications*, pp. 19–30, Amsterdam, the Netherlands: Elsevier.

Tosa, T., Sato, T., Mori, T., Yamamoto, F., Takata, I., Nishida, Y. and Chibata, I. (1979) Immobilization of enzymes and microbial cells using carrageenan as matrix, *Biotechnology and Bioengineering*, 21, 1697–1709.

Wang, T., Gu, Q., Zhao, J., Mei, J., Shao, M., Pan, Y., Zhang, J., Wu, H, Zhang, Z. and Liu F. (2015) Calcium alginate enhances wound healing by up-regulating the ratio of collagen types I/III in diabetic rats. *International Journal of Clinical Experimental Pathology* 8, 6636–6645.

Yabuki, M., Uchiyama, A., Suzuki, K., Ando, A. and Fujii, T. (1988) Purification and properties of chitosanase from Bacillus circulans MH-K1. *Journal of General Applied Microbiology* 34:255–270.

Some Bacterial and Synthetic Polysaccharides

6

The polysaccharides made by bacteria are quite different in function from those of higher plants. They are secreted from the cell to form a layer over the surface of the organism, often of substantial depth in comparison with the cell dimensions, which is believed to have several functions. Because of their position, they are described as exocellular or 'capsular' polysaccharides, to distinguish them from any polysaccharides that might be found within the cell (Figure 6.1). The functions are thought to be mainly protective either as a general physical barrier preventing access of harmful substances or more specifically as a way of binding and neutralizing bacteriophage. In appropriate environments, they may prevent dehydration. They may also prevent phagocytosis by other microorganisms or the cells of the immune system. The capsular polysaccharides are often highly immunogenic and may have evolved their unusual diversity as a way of avoiding antibody responses.

They also have a role in adhesion and penetration of the host, and since in the case of plants this will involve interaction with polysaccharide structures in the cell walls there are clearly possibilities of specific interactions. Plant lectins (glycoproteins) which have specific binding properties with respect to carbohydrate structures may play a part in this and general defences of plants against bacterial infection. Lectins are thought to be part of the plants' defence against insect and animal depredations and are known to bind to sites in the gut; their presence predominantly in the seeds is consistent with this, but they may also have a role against bacteria and fungi.

The secreted polysaccharides can be involved in pathogenicity. *Pseudomonas aeruginosa*, commonly found in respiratory tract infections, produces alginate which contributes to blockage in the respiratory tract, which leads to further infection (Massengale et al., 2000), while similar blockage of phloem in plants has been described (Griffin et al., 1996). On the positive side, these and other microbial polysaccharides provide great opportunity for biotechnological use.

6.1 BIOTECHNOLOGY

Bacterial products are very important in most aspects of polysaccharide biotechnology (Morris and Harding, 2009). They range from bulk materials such as xanthan, priced at about $14/kg, which are used

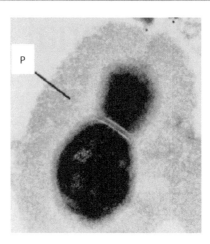

FIGURE 6.1   Exocellular or 'capsular' polysaccharide layer (labelled P) from *Streptococcus pneumoniae*. (Reproduced from Skov-Sørensen, U.B. et al., *Infection and Immunity*, 56, 1890–1896, 1988. With permission.)

mainly in food applications, to cyclic dextrans, valued at about $50/kg, which are used in high-value applications in research and pharmaceuticals.

Bacteria are known which produce nearly all the major plant polysaccharides such as glucans, alginate-like materials and even cellulose – though apparently not pectins – as well as the complex bacterium-specific materials. Genetic manipulation of bacteria has been studied for longer and is in general much easier than for higher organisms, so they are an obvious target both for manipulation of biosynthetic pathways and for the expression of heterologous genes to produce especially desirable enzymes. Polysaccharides are not of course under the direct control of genetic material in the way that enzymes are, and must be approached indirectly by manipulating the biosynthetic or degradation pathways by way of the enzymes responsible. We have already, in Chapter 2, seen one example of this, in the control of pectin degradation in higher plants, but there would seem to be great scope for this approach in bacteria as well. In addition, since there are now fermentation processes involving the large-scale growth of bacteria, all the advances in this that can be achieved by biotechnology will apply.

There are limitations. It is unlikely that large-scale fermentation could ever compete with existing processes for pectin manufacture, for example, since processors currently use waste by-products from other processes as their raw material. This is quite apart from the inherent costs of large-scale fermenters in comparison with simple extraction processes. It follows that bacterial products must have particularly useful properties in order to justify their probable greater cost. As we shall see there are a number of such products that can meet this criterion.

6.2 BACTERIAL ALGINATES

Although commercial alginates derive from algal sources (Chapter 5), there is a large potential for producing 'tailor-made' alginates from bacterial sources, especially if advantage is taken of the genetic tools for controlling the production of the enzymes that are responsible for the synthesis and epimerization (conversion of D-mannuronic, 'M', to L-guluronic residues, 'G') of the polymeric alginate chain (Hay et al., 2013). There appears to be greater structural diversity (poly-M, poly-G and poly-MG residues) and our understanding of the genes and the enzyme gene products is much greater for bacterial alginate production compared with the case for seaweed.

The main alginate-producing bacteria that have been studied are *P. aeruginosa* and *Azotobacter vinelandii*. *P. aeruginosa* has been the subject of particular attention because of its association with respiratory disease and is found in patients suffering from cystic fibrosis. *A. vinelandii* appears to be the most promising in terms of industrial production because of its stable output of alginate. *P. aeruginosa* alginate has no poly-G residues (and hence has a low G-content) whereas *A. vinelandii* can, like seaweed alginate, possess all three block sequences (poly-M, poly-G and poly-MG residues). However, all bacterial alginates are *O*-acetylated to varying degrees: acetylated mannuronic acids cannot be converted by epimerases from M to G.

6.2.1 ALGINATE STRUCTURAL ENZYMES

Starting from fructose-6-phosphate, the following enzymes are involved according to the scheme of Figure 6.2: hexokinase, phosphomannose isomerase, D-mannose-1-phosphate guanyl transferase, guanosine diphosphate (GDP) mannose dehydrogenase, transferase, acetyl transferase and mannuronan C-5-epimerase. Such schemes have been worked out for *A. vinelandii* (Pindar and Bucke, 1975), *P. aeruginosa* (May and Chakrabarty, 1994) and for the seaweed *Fucus garnweri*. Key structural genes have been identified in a gene cluster for *P. aeruginosa* (see May and Chakrabarty, 1994; Ertesvag et al., 1996) (Figure 6.3) with algG encoding for the epimerase and algF for the acetylase.

6.2.2 POSSIBILITIES

As Ertesvag et al. (1996) have reported, it may be possible to treat alginate with epimerases to increase the poly-G content, producing a firmer polymer. In principle, this could be done *in vivo* by the incorporation and expression of genes from a plasmid in an alginate-producing bacterium. The degree of acetylation is another control point since, as we have already mentioned, an acetylated mannuronic acid residue is immune from epimerization.

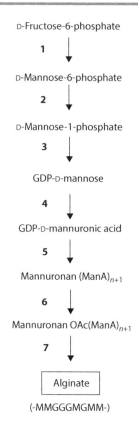

D-Fructose-6-phosphate

1

D-Mannose-6-phosphate

2

D-Mannose-1-phosphate

3

GDP-D-mannose

4

GDP-D-mannuronic acid

5

Mannuronan $(ManA)_{n+1}$

6

Mannuronan OAc$(ManA)_{n+1}$

7

Alginate

(-MMGGGMGMM-)

FIGURE 6.2 Alginate biosynthesis in *Azotobacter vinelandii* and the principal enzymes involved: 1: hexokinase; 2: phosphomannose isomerase; 3: D-mannose-1-phosphate guanyl transferase; 4: guanosine diphosphate mannose dehydrogenase; 5: transferase (*n* is the number of uronic acid residues in the polymer chain, that is, the degree of polymerization); 6: acetyl transferase; 7: mannuronan C-5-epimerase. Similar pathways for alginate production have been reported in *Fucus gardneri* (a brown alga) and *Pseudomonas aeruginosa*. (Reprinted from Pindar, D.E. and Bucke, C., *Biochemical Journal*, 152, 617–622, 1975.)

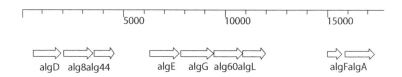

FIGURE 6.3 Gene cluster organization encoding most of the alginate structural enzyme genes in *P. aeruginosa*. Scale is in base pairs. (Reprinted from Ertesvag, H. et al., *Carbohydrates in Europe*, 14, 14–18, 1996 and May, T.B. and Chakrabarty, A.M., *Trends in Microbiology*, 2, 151–157, 1994, with permission from Elsevier.)

6.3 XANTHAN

Xanthan (or, as commonly called, 'xanthan gum') at \$14/kg falls into a similar price range to other gums and exudates and has similar functional properties. It is made by the organism *Xanthomonas campestris* (which in its wild existence is responsible for cabbage blight) grown largely on glucose, itself derived from maize starch, which it converts with high efficiency (80%) to the xanthan gum structure. Typically for a fermentation product, the raw material costs are a small part of the total, which are mainly to do with recovering the gum from the culture medium. Production is around 9000 tonnes a year. It has a β(1→4)-linked glucan main chain with alternating residues substituted on the 3-position with a trisaccharide chain containing two mannose and one glucuronic acid residue. It is thus a charged polymer. Some of the mannose residues may also carry acetyl groups. It is useful because it forms relatively rigid-rod-like structures in solution at ambient temperatures, though they convert to a more flexible conformation on heating. These rods are able to align themselves – like agarose and the κ- and ι-carrageenans – with the unsubstituted regions of galactomannans, such as guar and its derivatives and locust bean gum, to produce fairly rigid mixed gels with applications in food manufacture (see also Sections 2.4, 5.2 and 5.3).

The primary structure (as worked out by sequential degradation and methylation analysis) consists of the same backbone as cellulose:

$$...→4)\ β\text{-}\text{D-}\text{Glc}p\ (l→...$$

with the substitution at C3 of every alternate Glc*p* residue by the negatively charged trisaccharide

β-D-Man*p*-(l→4) β-D-Glc*p*A-(1→2) α-D-Man*p*-6-*O*-acetyl (1→3)

with some of the terminal β-D-Man*p* residues substituted at positions C4 and C6 by pyruvate ($CH_3.CO.COO^-$). Thus, xanthan (except under highly acidic conditions) is a highly charged polyanion. Figure 6.4 shows the Haworth projection of a (pyruvylated) alternate repeat section of xanthan.

It is not only charge repulsion effects that give xanthan an extended rigid-rod conformation, that is a 'zone A' (cf. Section 1.1.7) polysaccharide: x-ray fibre diffraction studies (Millane et al., 1989) have shown that, like the marine polysaccharides κ- and ι-carrageenan along with agarose, xanthan exists in 'ordered form' as a helix although there is still uncertainty as to whether this helix is a coaxial duplex or just a single chain one (with the projecting trisaccharides folded back along the axis of the helix giving effectively a chain rigidity similar to that of a double helix). Using a combination of electron microscopy (contour chain length, L) and light scattering (weight average molecular weight, M_w), Stokke and co-workers evaluated a mass per unit length, M_L of (1950 ± 250) Da nm^{-1} (Stokke et al., 1989; Kitamura et al., 1991). From x-ray fibre diffraction, a corresponding value of 950 Da nm^{-1} was obtained, and so a duplex or double-helical structure

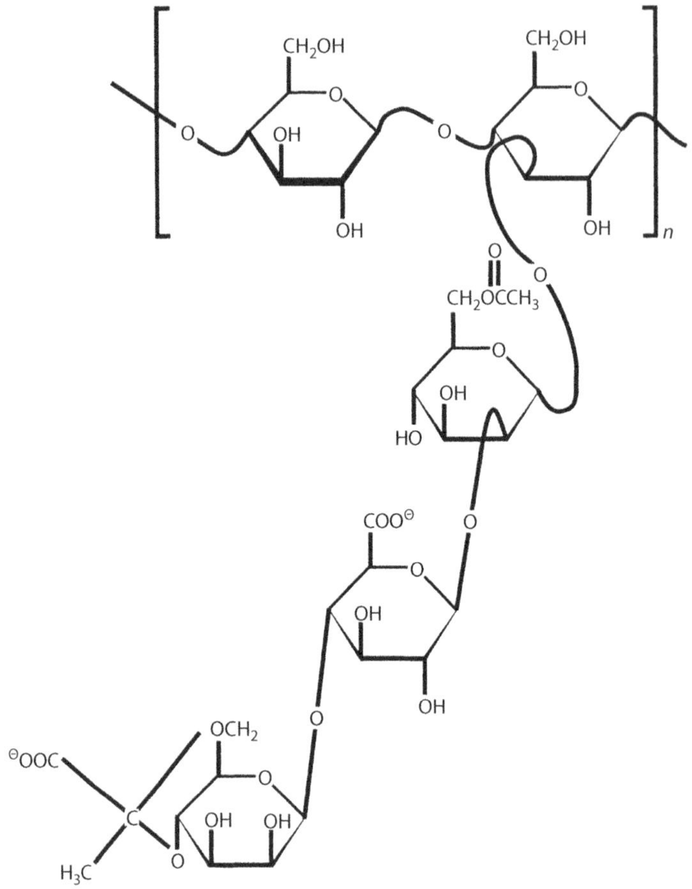

FIGURE 6.4 Xanthan repeat units showing a trisaccharide side chain with pyruvated end mannose unit. Not all the terminal side chain residues are pyruvated.

was inferred in solution. Further studies showed some strand separation and that the duplex structure was more rigid compared with regions that had been partly strand-separated. High temperatures will reversibly melt this helix to give a somewhat more flexible (zone C or D) structure.

6.3.1 EXTRACTION AND PRODUCTION

The primary laboratory and commercial fermentation medium – as has been known for over 50 years – for *X. campestris* growth and xanthan production is a phosphate-buffered (pH ~7) broth containing D-glucose (30 g L^{-1}) (or sucrose, starch, hydrolysed starch), NH$_4$Cl, MgSO$_4$, trace salts and 5 g L^{-1} casein (or soybean) hydrolysate, and the fermentation process takes place aerobically at a temperature of ~28°C. Xanthan production is further stimulated by the presence of pyruvic, succinic or other organic acids. The xanthan produced in this way is very similar to the xanthan produced naturally by the microbe living on a cabbage. In the commercial process, the oxygen uptake from the broth is controlled to a rate of 1 mmol L^{-1} min^{-1}. Treated in this way, the bacterium is an extremely efficient enzyme mini-factory converting >70% of the substrate (D-glucose or related) to polymeric xanthan. The bacterium having done its work is then removed in a rather undignified way by centrifugation and the xanthan precipitated with methanol or 2-propanol at 50% weight concentration. The xanthan slurry is then dried and milled for use. The original commercial producer of xanthan was Kelco Ltd. (now MSD-Kelco) and together with other suppliers the annual worldwide production is now over 10,000 tonnes.

The biosynthetic process by the bacterium worked out by Sutherland (1989) and later confirmed by others follows the same basic pattern proposed for other microbial polysaccharides:

1. Substrate uptake
2. Substrate metabolism
3. Polymerization
4. Modification and extrusion

This involves lipid carriers, although there is still uncertainty over their precise role and how the whole process is controlled.

6.3.2 PROPERTIES

Table 6.1 summarizes the fundamental properties of a popular commercial xanthan. Analytical ultracentrifugation, light scattering and viscometric studies using Mark–Houwink, worm-like coil and Wales–van Holde types of analysis are all consistent with a rod conformation in solution. The weight average M_w of around 6 million daltons for Keltrol-xanthan (Dhami et al., 1995) means that this, along with other commercial

---

TABLE 6.1   Fundamental Properties of a Commercial Xanthan: Keltrol (Nutrasweet-Kelco Ltd, USA)

Property

Polymer type	Polyanionic; complex repeat
Conformation type	Zone A/B (Extra-rigid/rigid-rod)
Persistence length, L_p	100–150 nm
Solubility	Cold or hot aqueous systems
Molecular weight	$(5.9 \pm 1.1) \times 10^6$ Da
Intrinsic viscosity [η]	~7500 mL g^{-1}
Sedimentation coefficient, $s_{20,w}$	(13.0 ± 0.3)S
Wales–van Holde ratio, $k_s/[\eta]$	0.3 ± 0.1
Mark–Houwink a coefficient	1.2
Gelation properties	Transient weak gels
Key sites for biomodification	C3 trisaccharide side chains (particularly extent of acetylation of terminal β-D-Man$p$); helical backbone – synergistic interactions with galactomannans

xanthans, is one of the largest of the aqueous soluble polysaccharides. Solutions are correspondingly extremely viscous. The intrinsic viscosity of ~7500 mL g^{-1} is one of the highest known for a polysaccharide, and the dilution solution concentration limit via the Launay C^* parameter (~3.3/[η], Section 1.1.8) is very low (~0.5 mg mL^{-1}). The very high viscosity at low concentrations (e.g. at 5 mg mL^{-1} a viscosity of ~1000 cP has been observed at room temperature) makes it ideal as a thickening and suspending agent. By itself xanthan, however, only forms transient weak gels since the junction zones are weaker than in those used for networking in carrageenan and agarose (Sections 5.2 and 5.3).

6.3.3 MODIFICATION

The key sites for modification are the first and terminal mannose residues on the trisaccharide side chains (particularly extent of acetylation of the first and pyruvalation of the latter) and the helical backbone by forming non-covalent interactions with galactomannans: these interactions appear to be also affected by the substitutions in the side chains. As to the trisaccharide side chains themselves, there are two approaches for alteration or control: one is changing the physiological conditions of fermentation; the other is the use of different pathovars or strains of *X. campestris*. The pathovars *pv phaseoli* and *oryzae* yield virtually acetyl-free or pyruvate-free xanthan, respectively.

The other approach (see e.g. Vanderslice et al., 1989) has been to look at the genetics of the enzymes controlling the biosynthetic pathway and to attempt to produce and isolate in sufficient quantity genetic mutants deficient or defective in one or more of these enzymes to give 'polytetramer' (i.e. lacking in the terminal mannosyl or pyruvalated mannose group) and 'polytrimer' (lacking in addition the adjacent glucuronic acid residue). Although yields of 50% polytetramer have been found, corresponding attempts for polytrimer and other xanthan variants have thus far been disappointing.

There has been considerable interest in improving the weak gelation characteristics of xanthan by inclusion of galactomannans such as locust bean gum into mixtures (see e.g. Dea et al., 1977). We have already described the use of synergistic interactions between galactomannans with carrageenan and agarose in Chapters 2 and 5. The results are more marked with xanthan whose own gels are weak and transient, but made in the presence of guar or locust bean gum give stronger gels with an optimum mixing ratio ~50:50 by weight. Deacetylating the xanthan side chains seems to enhance these synergistic interactions.

6.3.4 USES OF XANTHAN

Xanthan has been food grade approved by the US Food and Drug Administration for nearly 50 years. This makes it not only attractive as a food product (see e.g. Sanderson, 1982) but also for use in packaging material in contact with food and in pharmaceutical and biomedical applications that involve ingestion. Its uses chiefly derive from its solubility in hot and cold water and its very high thickening and suspending potential which in turn derives from the very high viscosity of its suspensions. Despite the high viscosity, xanthan suspensions exhibit high shear thinning which means they also flow easily (i.e. good pourability).

6.3.4.1 FOOD APPLICATIONS

Besides high viscosity, thickening and suspending ability, xanthan suspensions have high acid stability. This makes them highly popular in sauces, syrups, toppings and salad dressings. In drinks, the addition of xanthan together with carboxymethylcellulose adds 'body' to the liquid and assists with uniform distribution of fruit pulp and so on. It is also used to add body to dairy products. The high freeze–thaw stability of xanthan suspensions makes them particularly attractive for the frozen food industry. The high suspending and stability properties of xanthan is also taken advantage of by the animal feed industry for transporting liquid feeds with added vitamins and other supplements that would otherwise sediment out with transport or storage time.

6.3.4.2 PHARMACEUTICAL AND COSMETIC USES: MICROENCAPSULATION

Xanthan has been added to the list of hydrophilic matrix carriers, along with chitosan, cellulose ethers, modified starches and scleroglucan (Melia, 1991). Tablets containing 5% xanthan gum were shown under low shear conditions to give successful controlled release of acetaminophen into stomach fluid, and tablets containing 20% xanthan successfully carried a high loading (50%) of the drug theophylline.

The high suspension stability is made use of in pharmaceutical cream formulations and in barium sulphate preparations. This high cream stability is taken advantage of also by the cosmetic industry (Grollier, 1984), including, as with carrageenan (Section 5.2), toothpaste technology where the toothpaste will hold its ingredients (high viscosity) and then easily brush onto and off the teeth (high shear thinning). Uniform pigment dispersal, along with other ingredients and long-term stability, makes xanthan a good base for shampoos.

6.3.4.3 OIL INDUSTRY

Xanthans are used in a number of aspects of well-bore technology such as (1) a proppant into fissures for rock fracturation, (2) the suspension and removal of debris from the bore area even in the presence of harsh environments (seawater included) and pipelines and (3) enhancing oil recovery using the technique of 'polymer flooding'.

6.3.4.4 WALLPAPER ADHESIVES

The high stability of suspensions makes xanthan an ideal base for suspending adhesive agents for wallpapers.

Like alginates (Section 5.1), the suspension-stabilizing property of xanthans makes them ideal for producing sharp prints from dyes with a minimum risk of running – for application to textiles and (in conjunction with guar) carpets.

6.4 PULLULAN

Pullulan is an α-glucan made up from maltotriose units linked by α(1→6) bonds. It is obtained from *Aureobasidium pullulans* and is hydrolyzed by pullulanase to yield maltotriose. It is not attacked by digestive enzymes of the human gut and is used to form films. Production is now substantial and has found particular application in formulating snack foods in Japan based on cod roe, powdered cheese and as a packaging film for ham.

It is water soluble, with molecular weights in the range 5 - 900 kDa with straight unbranched chains, and it behaves as a random coil according to a combination of sedimentation coefficients and intrinsic viscosity measurements. Similarly, light scattering and sedimentation equilibrium measurements show that pullulans above about 50 kDa molecular weight behave to a close approximation like a classical random coil (Kawahara et al., 1984). Pullulan has been proposed as a 'standard polysaccharide' in the sense that it is so near to random-coil behaviour, and it is readily obtainable in very reproducible form and so it could be used for comparative tests with other polysaccharides. The idea is presumably that deviation from pullulan-type behaviour must imply a more complex structure. It might also be used to standardize instruments. The solutions in water are liable to bacterial attack with rapid degradation of the molecular weight and preservatives such as azide may be needed.

6.5 GELLAN, XM6 AND CURDLAN

Inspired by the example of xanthan, other bacterial polysaccharides with useful properties have been developed, in the case of gellan as the result of a systematic search for a polysaccharide of the required properties followed by the identification of the organism (Morris, 1987).

Gellan is obtained from cultures of *Pseudomonas elodea*, which is found growing on the *elodea* plant. It is isolated by ethanol precipitation from the culture medium and may be partially deacetylated by alkali treatment. It has a linear structure with a repeat unit of a tetrasaccharide (see Figure 6.5) each with one carboxyl group and in the native state one acetyl group. It is therefore sensitive to calcium levels but has theological properties similar to those of xanthan which has a similar charge density. It was clearly intended to be a xanthan competitor, though before permission for food use was obtained it was promoted as an agar substitute, particularly for use in growth media.

The XM6 repeat unit structure is also shown in Figure 6.5 and is a polysaccharide made by an *Enterobacter* discovered at Edinburgh

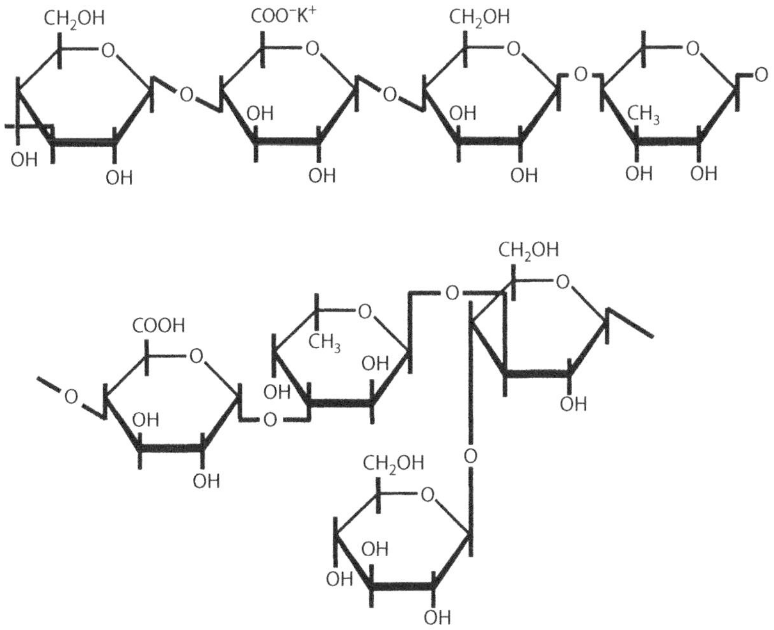

FIGURE 6.5   Repeat units, at the top for gellan in deacetylated form and below for the XM6 polysaccharide.

University. It lacks acetyl groups but is classed along with xanthan and gellan and can be induced to gel by adding calcium ions. It is included here as a representative of many similar bacterial polysaccharides, all of which have interesting and potentially useful rheological properties. It is not commercially available and not approved for food use, and probably never will be.

Curdlan is one of two polysaccharides produced by *Alcaligines faecalis* with a repeat unit of three β(1→3)-linked D-glucose residues, and unlike the examples earlier is not charged. Production is now from *Agrobacterium* mutants that make only curdlan in high yield and is mostly confined to Japan. Curdlan suspensions form gel on heating – apparently irreversibly. There are a number of potential applications for it in the food industry, mostly as a replacer for existing gums, and it may be used in Japan in some products. Novel products based on its heat-set ability may also appear.

6.6 DEXTRANS

Bacterial dextrans are produced in substantial quantities by *Leuconostoc mesenteroides* and are familiar to laboratory workers as the basis for cross-linked dextran beads used in gel filtration columns. They are mostly α(1→6)-D-glucopyranosyl polymers with molecular weights up to ~1 × 10^6 Da more or less branched through (1→2), (1→3) or (1→4) links. In most cases, the length of the side chains is short and the branched residues vary between 5% and 33%. The major commercial dextran is about 95% (1→6) linked and 5% (1→3) linked, and it is made by selected strains of *Leuconostoc*. After ethanol precipitation from the culture medium, acid hydrolysis is used to reduce the overall molecular weight, though fungal dextranases can also be used. The product with an average molecular weight of about 60,000 Da is used in medicine as a blood extender, while fractions of defined molecular weights (e.g. the Pharmacia (Pharmacosmos) 'T-' series, where 'T500 Dextran' would stand for dextran of weight average molecular weight 500 kDa) are familiar in laboratories and to some extent, like the pullulan 'P-' series, serve as polysaccharide standards in molecular weight calibrations. They are also used as part of incompatible phase separation systems, usually with polyethylene glycol (PEG). 'Blue dextran' is a well-known marker for the void volume for gel filtration studies.

Dextrans have found very wide application in laboratory work because they are particularly free from positive interactions with proteins. The interactions can be almost entirely characterized as co-exclusion. This has found application, as already noted, in gel filtration media, but cross-linked dextran gels show other effects. For example, they swell and shrink in a way related to the osmotic pressure of the solvent system and can be used to make miniature osmometers: the reader is referred to an old but notable study by Ogston and Silpananta (1970).

6.7 CYCLODEXTRINS

Cyclodextrins have attracted more interest than any other bacterial polysaccharide because of their unique structures. The interest has been more scientific than industrial, since xanthans are much more important in commercial terms, but there always seems to be a potential application of the cyclodextrins on the horizon which would lead to a major expansion in demand for them. Formerly called Schardinger dextrins after their discoverer, and still called cycloamyloses and cyclomaltoses here and there, they are cyclic molecules formed by glucopyranose linked by α-D-$(1{\rightarrow}4)$-linked bonds. They contain six, seven, eight or nine glucopyranose residues. It is impossible to form cyclic structures with less than six residues, while those with more than nine have so far been found to have at most a nine-membered ring with other glucose residues attached as branches through $(1{\rightarrow}6)$ links. The structures (for a seven-membered ring, $cGlc_7$) are illustrated in Figure 6.6. The nine-membered ring is uncommon and found only as a minor by-product of enzymatic synthesis of the others. The six-, seven- and eight-membered rings are sometimes called the α-, β- and γ-cyclodextrins, abbreviated as $cGlc_6$, $cGlc_7$ and $cGlc_8$.

They are water soluble and easily crystallized and their structures have been determined. The great interest in them is because the cavity is of a size to include many small molecules of interest and is also fairly hydrophobic in character. The outside of the toroidal cone is full of hydrophilic hydroxyl groups. As might be expected, most of the potential applications involve inserting smaller molecules into the cavity. The dimensions are roughly those of a cylinder with length of 0.7 nm and diameter of 0.45 nm for the six-membered ring and 0.85 nm for the

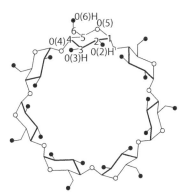

FIGURE 6.6   Diagrammatic structure of cGlc$_7$, a cyclodextrin with a seven-residue ring. (Bender, H.: Production, characterisation and application of cyclodextrins. *Advances in Biotechnological Processes*, 1986, 31–71. Copyright Wiley-VCH Verlag GmbH & Co. KGaA. Reproduced with permission.)

eight-residue ring. Thus, they can accommodate an aromatic ring like benzene, as well as alkyl chains.

6.7.1 BIOSYNTHESIS

The enzymes capable of synthesizing cyclic dextrins are exclusively bacterial in origin.

Bacillus macerans is the best-known bacterium producing an amylase with glycosyl transferase activity. Seven other species are known, including *Bacillus stearothermophilus*, *Bacillus circulans* and *Klebsiella pneumoniae*. The enzyme is induced when they are grown on starch, and the mechanism is believed to involve fission at the non-reducing ends of the chains, followed by cyclization as suggested in Figure 6.7. They can also act as general transferases. This is a variation on the usual amylase mechanism where the group which becomes attached to the enzyme is released by hydrolysis, and the glucotransferases show strong homology with the α-amylase family and clearly belong to it. The crystallographic structures of the *B. circulans* and *B. stearothermophilus* enzymes are available. The enzyme can act in the vicinity of branch points and can produce branched products of the type shown in Figure 6.7.

There is some doubt about the formation of a covalent link between the C1 carbon and the enzyme active centre aspartate residue. Evidence from pancreatic amylase supports the idea but crystallographic data suggest that the barley amylase does not. The distance between the aspartate nucleophile and the C1 carbon is too great. The point is not crucial since there are well-established binding sites for the polysaccharide chain on the enzyme and the exact location of these must play a role in the cyclization reaction.

Different bacteria make enzymes with slightly different specificities which have been exploited to make different cyclodextrins. Thus that from *Bacillus 38-2* produces mainly the cGlc$_7$ while *B. macerans* and

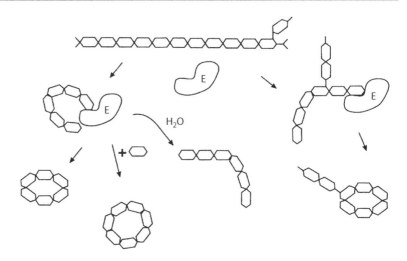

FIGURE 6.7 Mode of action of cyclodextrin synthases. They act as amylases with general transferase activity but can also cyclize the product. Branched derivatives can also form as shown.

Klebsiella are used to make the cGlc$_6$ dextrins. However, it is possible to reduce the seven-membered ring to the six-membered ring by using a mixture with glucose and the transferase enzyme even where it is of the c7 type. There are no enzymes with a cGlc$_8$ preference which occurs as a minor (5%) by-product of the others.

6.7.2 LARGE-SCALE PRODUCTION

More important from the point of view of production methods is the tendency of the enzyme to make linear fission products under some conditions. Once the enzyme has split the main chain, the fate of the groups attached to it depends on the chances of collision with potential receptors and the relative rate of reaction. Thus, high substrate concentrations will lead to a better chance of transfer to another sugar residue rather than cyclization (or water) and there is an optimum substrate concentration to produce cyclic fragments. At less than 1% starch, 85% of amylose and 65% of amylopectin are cyclized, while above 5% starch only 35%–50% cyclization can be obtained. Adding other enzymes such as pullulanases or isoamylase, a debranching enzyme, will also shift the ratios. In practice, a compromise between the yield and the inconvenience of low concentrations of starch in the form of large vessels and costly water removal leads to a level of about 15% starch although a limited prior hydrolysis to avoid retrogradation is desirable.

Total worldwide production is now in excess of 100 tonnes a year, and it is exclusively by batch methods in stirred tanks. Most of it is the cGlc$_7$ at a cost of about $10/kg, but processes for the six- and eight-member rings also exist, though at much higher prices. Potato starch is the usual substrate, but maize starch has also been used. The subject has been reviewed by Bender (1986), who distinguishes two kinds of process. Figure 6.8 gives an outline of two methods for the manufacture of cGlc$_6$. One of them uses decanol as an aid, since it forms a complex with the cyclic dextrin with relatively low solubility. This both pulls the reaction over towards the cyclized product and aids its isolation. It is removed by steam distillation followed by crystallization of the cGlc$_6$. In the other, complexing agents are not used and the purification is by cation exchange resins and crystallization.

Processes for making cGlc$_7$ are easier because it has a lower solubility than the six-member ring. A process similar in principle to those shown in Figure 6.8 is shown in Figure 6.9. This time toluene is used as the complexing agent. Unlike decanol this will not so easily find food use approval.

The eight-membered ring compound of the series is isolated from the supernatants and mother liquors of the six- and seven-ring processes. It is purified by chromatographic methods. One report suggests that digestion with the *Klebsiella* enzyme in the presence of sodium acetate increases the proportion of cGlc$_8$. This particular cyclodextrin forms a complex with bromobenzene and the yield can be increased by including it.

Decanol complex process
Soluble potato starch
20% in water, 80°C, 5 minutes
↓
Cycloglucan transferase
ex *Klebsiella*. Decanol (v/v),
40°C, 24 hours
↓
α-amylase, 65°C, 6 hours
↓
Decanol complex precipitates
↓
Water suspension, steam distillation
to remove decanol
↓
Filtration, concentration and crystallization
of cGlc$_6$

Non-complexing process
Potato starch, 6% in water
120°C, 30 minutes
↓
Cycloglucan transferase ex *B. macerans*
40°C, 40 hours
↓
120°C, 30 minutes, adjust to pH 5
Glucoamylase, amylase
40°C, 30 hours
↓
Ion exchange, activated charcoal, concentrate
↓
Cation exchange chromatography
↓
Concentrate and crystallize cGlc$_6$

FIGURE 6.8 Two schemes for the production of cGlc$_6$. (Bender, H.: Production, characterisation and application of cyclodextrins. *Advances in Biotechnological Processes*. 1986. 6. 31–71. Copyright Wiley-VCH Verlag GmbH & Co. KGaA. Reproduced with permission.)

Toluene complex process
Maize starch, 28% in water
80°C, 10 minutes, 120°C, 30 minutes
B. subtilis α-amylase
↓
B. macerans cycloglucan transferase
50°C, 30 minutes
Toluene 5% (v/v)
45°C, 105 hours
↓
Toluene distillation
50% water, 50% ethanol
↓
Supernatant, extra toluene
Collect complex
↓
Distill toluene
Crystallize, recrystallize,
active charcoal
↓
Final crystallization of cGlc$_7$

FIGURE 6.9   A scheme for the production of cGlc$_7$. (Bender, H.: Production, characterisation and application of cyclodextrins. *Advances in Biotechnological Processes*. 1986. 6. 31–71. Copyright Wiley-VCH Verlag GmbH & Co. KGaA. Reproduced with permission.)

Traces of cGlc$_6$-bearing branches are formed and have been isolated and crystallized. Although they have slightly different complexing abilities and higher solubilities, they have not been seriously considered for use because of their scarcity.

6.7.3 MANUFACTURE OF CYCLOGLUCOTRANSFERASES

The enzyme from *Bacillus 38-2* is isolated partially purified from culture medium by an unusual method. This involves ethanol precipitation at −20°C, drying and resolution in water, complexing of the enzyme with maize starch, followed by elution with phosphate buffer and chromatography on gel filtration and diethylaminoethyl (DEAE)-cellulose. As a final stage, preparative electrophoresis was used. This cannot be considered to be a serious industrial enzyme isolation and is much nearer to a small laboratory isolation procedure. Another method appears to be better suited to larger scale operation. This involves the use of an immobilized cGlc$_6$ column directly on the culture medium, followed by elution with the cyclodextrin. This gave a 100-fold purification and a 90% yield. This is clearly the best method and the one to consider first when making these enzymes. Some selectivity can be obtained with cGlc$_7$ columns for enzymes favouring that product.

Surprisingly, considering the nature of the reaction, a transferase immobilized on a vinylpyridine polymer matrix was effective, though it lost 75% of its activity, and apparently requires a very long spacer arm. It could be repeatedly used over 2 weeks at least. Other transferases have also been successfully immobilized.

6.7.4 APPLICATIONS

6.7.4.1 AMYLASE INHIBITION

Cyclodextrins are competitive inhibitors of α- and β-amylases, potato phosphorylase and pullulanase. They are similar therefore to a number of other amylase inhibitors that have been found and which are mainly small carbohydrate molecules that interfere with the active site of the enzyme. A different class of non-competitive protein amylase inhibitors is also known to exist.

This property immediately raises two questions. First, do the cyclic dextrins present any hazards, perhaps as part of a foodstuff, or possibly by inhalation of a powder in a factory environment? Second, is there any possibility of using this property as a component of slimming aids? Feeding trials with rats suggest that $cGlc_7$ is relatively digestible compared with $cGlc_6$ which has been considered as a possible component of a low-calorie diet. It can be regarded as a component of the fibre, on par with the indigestible part of starch, but considering its high cost and the fact that starch can also provide adequate amounts it is unlikely ever to find applications of this kind. Toxicity trials showed an LD50 of 12 g kg^{-1} in rats for oral intake, which is high, and values around 1 g kg^{-1} for intravenous injection for both $cGlc_6$ and $cGlc_7$. There are unlikely to be handling problems.

6.7.4.2 ARTIFICIAL ENZYMES

The idea of artificial enzymes, designed to have a specific catalytic activity, is rather vague, since it might be thought to apply strictly to peptide chains synthesized *de novo*, albeit with the aid of synthetic cDNA and the cells' protein synthesis system. When this term was introduced this was not feasible, though it is now, and instead it came to mean some sort of polymer, not necessarily of biological origin since polyacrylamide has been involved, that is able to show catalytic activity similar to that shown by *bona fide* enzymes of biological origin. In all discussions, cyclic dextrins play a prominent part.

If the effectiveness of the 'enzymatic activity' is assessed by comparing the rate of hydrolysis in the presence and absence of the cyclodextrin, then the largest effect is on the hydrolysis of an ester group on E-3-carboxylene 1,2-ferrecene cyclohexane *p*-nitrophenyl ester which is accelerated ~3,200,000 times. More typical is the hydrolysis of a *t*-butylphenylacetyl ester, which is accelerated ~250 times. There is a long list of esters, amides and organophosphates which are apparently able to hydrolyze more rapidly via the formation of a covalent link to the cyclodextrin. They show chiral specificity and other typical kinetic enzyme properties such as competitive inhibition and are indeed convincing models for enzymes. The effects are not confined to hydrolase-like activity; decarboxylation and oxidation group migration have both been described. The mechanisms are closely similar to those of enzymes themselves. So far applications seem to be confined to laboratory synthesis.

6.7.4.3 INCLUSION COMPLEXES

The ability of cyclodextrins to enclose smaller molecules and effectively to sequester them from solution has led to a number of applications. One recent suggestion is as an aid to protein folding. Sodium dodecyl sulphate (SDS) interacts strongly with proteins, to form a complex with high solubility, but where the peptide chain is generally extended into a rod shape. Removing the SDS is difficult and the yield of correctly folded protein tends to be poor. When the complex of SDS and the enzymes carbonic anhydrase and citrate synthase was treated with $cGlc_7$, in both cases the SDS was sequestered and the enzymes folded to the correct active form in good yield. Since present methods involve tedious and lengthy dialysis procedures this could be useful in recovering badly folded proteins from heterologous expression systems. In an application such as this the high cost would be justified.

Formation of clathrates can be used to stabilize volatile and otherwise unstable materials. The complexes may also show enhanced solubility in water and a long list of drugs are known to be improved in their handling properties by the formation of inclusion complexes. Improved absorption can result from encapsulation so as to deliver materials to the appropriate part of the intestines. In food applications, flavours and fragrances are more effective when used in the form of complexes. Having said all this it is not easy to point to any particular application, though it is likely that most of the world production goes into a large number of small outlets in pharmaceuticals. One problem is that encapsulation effects can usually be achieved by other, probably cheaper means.

6.7.5 DERIVATIVES OF CYCLODEXTRINS

Cross linking the cyclodextrins yields some interesting gel filtration matrices. They can be used for chiral resolution and for the removal of substances such as *naringin* and *limonin* from fruit juice, where they give unwanted bitter flavours. Pairs of cyclodextrins linked by bridges have also been made, with modified complexing properties, but these are only laboratory curiosities so far. Of greater potential use are the cyclodextrins attached to an agarose matrix since these are useful for the isolation of amylases, for which a demand does exist. Modified cyclodextrins can be provided with metal-chelating groups, such as diaminoethane, and this extends the range of enzyme-like behaviour (Bender, 1986).

6.8 SCLEROGLUCAN AND SCHIZOPHYLLAN

A consideration of microbial polysaccharides would not be complete without at least a brief consideration of two fungal polysaccharides that are attracting increasing commercial interest: the weak-gelling scleroglucan and the schizophyllan systems. They are both large, neutral polysaccharides of weight average molecular weights, M_w, (as largely established by light scattering techniques) ~500 kDa, with a greater diversity being reported for scleroglucan. They both appear from x-ray diffraction to exist as hydrogen-bond stabilized triple helices, with resultant extra-rigid-rod like properties in solution: they have virtually the largest persistence lengths known for polysaccharides: ~150 nm for schizophyllan (see e.g. Yanaki et al., 1980) and 200 nm scleroglucan (Biver et al., 1986). The existence of the triple helix for scleroglucan has been further supported by electron microscopic/light scattering measurements of the mass per unit length, $M_L = (2100 \pm 200)$ Da nm^{-1} (Kitamura et al., 1994), along similar lines to those that supported the duplex model for xanthan as discussed earlier. The Mark–Houwink viscosity a parameter of 1.7 for schizophyllans of M_w <500 kDa is again amongst the highest known for a polysaccharide and is on the rigid-rod limit (Yanaki et al., 1980). For chains of M_w >500,000 Da, the a parameter falls to ~1.2 and corresponds to slightly more flexibility as the polymer length increases.

Chemically also they are very similar, with a backbone of repeating β(1→3)-linked glucose residues:

$$\ldots\rightarrow3)\ \beta\text{-}\text{D}\text{-Glc}p\text{-}(1\rightarrow\ldots$$

In scleroglucan, every third residue has a β(1→6)-linked D-glucose branch which protrudes from the triple helix. Stokke et al. (1993) have demonstrated using electron microscopy that certain denaturation–renaturation treatments cause the formation of interesting ring structures or 'macrocycles'. In common with other branched glucans they appear to stimulate an immune response against tumour cells (Kurachi et al., 1990) and, particularly scleroglucan, have been considered for use in cosmetics (as part of skin and hair products), for application in pesticides (to assist binding to foilage) and, along with xanthan and other polysaccharides, advantage can be taken of their high capacity to bind water and their high heat stability in oil well drilling fluids.

6.9 SYNTHETIC POLYSACCHARIDES: USE OF ENZYMES

While chemical synthesis of polysaccharides and oligosaccharides is possible in principle, synthetic strategies are lengthy, complicated and skilled-labour intensive. Generally, the main way of achieving this is likely to be by the judicious use of enzymes (Rastall and Bucke, 1992). The demand for synthetic oligo- and polysaccharides is stimulated mainly by experimental work. It is known that complex oligosaccharides on the surface of cells, usually in the form of conjugates with protein or lipid, are strongly antigenic. This is so whether the cells in question are bacterial (where gram-negative organisms have lipopolysaccharide determinants), erythrocytes (since blood groups depend on surface oligosaccharides for their specificity) or transformed tumour cells which have characteristic groups. In the last case, potential diagnostic methods could be developed if suitable ways of detecting these changes were available. It is also possible to target antibodies onto the sites in question leading to potential treatment methods.

Interactions between plants and bacteria such as the nitrogen fixers also depend on specific carbohydrate recognition sites. All these areas have stimulated a demand for oligosaccharides of known and often complex structure, in small amounts and where the materials would have high value. This is quite different from the demand for bulk supplies, such as the 9000 tonnes per annum of xanthan currently sold, and means that enzyme methods which would be simply not worth consideration can be used, or at least considered. Thus, all the known methods of polysaccharide biosynthesis, including those using nucleotide cofactors, have been investigated with this application in mind.

6.9.1 NUCLEOTIDE-DEPENDENT SYNTHASES

The way in which nucleotide-dependent synthesis occurs is well established. Two enzymes are involved, the nucleoside transferase which attaches the nucleotide to the sugar and a glycosyl transferase which then uses this high-energy intermediate to glycosylate an acceptor molecule, which it often does with very high specificity. It is this specificity that enables the directed synthesis of defined oligosaccharides. Many have been made and clearly within the limitations of the available specificities considerable control can be achieved. The oligosaccharides are relatively small and the method could not easily be extended to larger molecules. It is possible that chemical methods could be used to link these oligosaccharides to larger chains, but their main use is to see if they block determinant sites for which high molecular weights are not needed. Problems are that the enzymes are not readily available and of course many would be needed so as to exploit their specificity, and they are difficult to store and lose activity. These are all matters that might be improved by genetic engineering approaches, but it is uncertain that the size of the applications would justify it.

6.9.2 GLYCOSIDASE REVERSAL

A second possible approach is to use glycosidases in the reverse of their normal function. In fact, all glycosidases are potentially reversible and after the enzyme substrate complex has been formed can transfer the residue to either water or other receptors such as another sugar or to an alcohol. It is known for example that during the hydrolysis of lactose by lactases, in a commercial process, one of the problems is that the galactose residue can be transferred to a passing lactose to form a trisaccharide, and even higher polymers can be formed. Eventually, they will probably be broken down again, but this may be too slow for the process. They might also be insoluble or attach themselves to some other component.

In order to reverse hydrolysis, it is necessary to greatly increase the concentration of the monomer and also keep the water concentration as low as possible. The action of yeast α-glycosidase on mannose has been investigated. It was possible to get up to about four residues with a variety of acceptor sugars and a number of other studies using fungal enzymes reached a similar conclusion. Typical conditions were 83% w/w mannose, at 60°C with α-mannosidase, and only about 4% maltotriose was formed, with correspondingly less of the higher polymers. There have also been attempts to use organic solvents to keep the water level low, without dramatic effect.

One particularly interesting attempt made use of aqueous two-phase systems. An outline of the method is given in Figure 6.10 and, apart from its use in oligosaccharide synthesis, illustrates an application of bacterial dextrans in forming the two-phase system. It is based on the observation that α-mannosidase is heavily concentrated into the dextran-rich phase (this probably indicates a dextran binding site on the mannosidase; proteins in general do not concentrate into one phase in this way). The p-nitrophenol esters of mannose and galactose were the donors and acceptors, and the

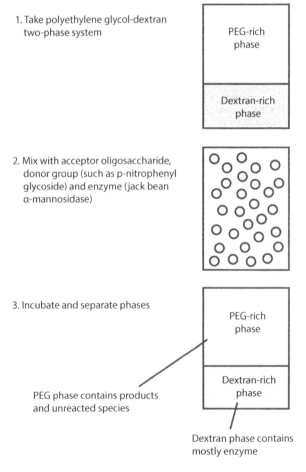

FIGURE 6.10   Use of a polyethylene glycol dextran two-phase system for the synthesis of oligosaccharides. The basic idea is that the α-mannosidase used is mainly contained within the dextran-rich phase, which can be repeatedly treated with reactants in the PEG-rich phase, with the loss of relatively little enzyme in each cycle.

main advantage of the system was efficient use of the enzyme. Other processes using immobilized enzymes have been described. The main problems with this method are the relatively low specificity of the transfers and the inherent difficulty of obtaining high degrees of polymerization. It may be a method for obtaining certain tri- and tetrasaccharides if a demand should develop, but is likely to be less useful than the third method of oligosaccharide synthesis using sucrose-dependent glycosyl transferases.

6.9.3 GLYCOSYL TRANSFERASES

These enzymes are probably the best prospect for directed polysaccharide synthesis, because they use the energy-rich bond of sucrose (or in some instances other disaccharides) to drive the addition of the residue to the growing chain. The point has already been made when discussing the synthesis of fructans, which are made *in vivo* by enzymes using this reaction (see Chapter 5).

The dextran sucrase of *L. mesenteroides* has been extensively investigated to find the scope of the acceptors capable of receiving the transferred group. The group transferred is the glucose half of sucrose, and the newly formed links are almost always α(1→6), but a range of receptors can operate. In addition, the newly formed oligosaccharide can itself act as a receptor so that quite long chains can be built up. This is certainly the case with fructans, where the initial receptor is sucrose.

Another transferase is found in *Streptococcus mutans*, an organism which produces an α(l→3)-linked glucan with some α(l→6) links. This is insoluble and may be involved in adhesion of the bacteria to teeth. They also produce a soluble dextran with α(l→6) links but also 27% α(1→3) links, which depends on a different transferase. Another enzyme from *L. mesenteroides* synthesizes an *alternan*, with alternating (1→3) and (1→6) linkages, once again using sucrose as the donor. Table 6.2 shows the result of substituting different receptors and indicates the range of sugars that can act in this capacity.

The cyclodextrin glucosyl transferases have already been mentioned earlier under cyclodextrin. They can, however, use a wide variety of sugars as receptors, and some that can and some that cannot are listed in Table 6.3. In one of the most promising uses of this enzyme, a maltohexose was transferred to a cellopentose to create an 11-residue chain. Clearly these enzymes can use quite large chains, but there seems to have been no systematic attempt so far to explore the limits of size. As indicated earlier, most of the demand for synthetic oligosaccharides so far is for relatively small ones, which can block determinant sites on cell surfaces, and there is no serious demand for, for example, 37-residue linear dextrans which can probably be obtained in other ways – with some difficulty.

6.9.4 APPLICATIONS IN ANIMAL FEEDS

Bacteria attach themselves to surfaces by interaction of surface lectins, usually glycoproteins, with the surface carbohydrates of the cells they are hoping to invade. This is particularly true of pathogens such as

TABLE 6.2   Acceptor Sugars for the Transfer of Residues from Sucrose by *Leuconostoc mesenteroides*

Acceptor Sugar	Yield[a]
D-Glucose	6.9
D-Mannose	2.2
D-Galactose	1.3
D-Fructose	4.8
L-Sorbose	0.4
D-Xylose	0.4
Methyl α-glucoside	23.4
Methyl β-glucoside	8.0
1,5-Anhydro-D-glucitol	9.1
α,α-Trehalose	0.0
α,β-Trehalose	ND
β,β-Trehalose	ND
Maltose	48.0
Nigerose	20.5
Cellobiose	5.4
Isomaltose	43.0
Isomaltulose	ND
Gentiobiose	ND
Melibiose	ND
Sophorose	ND
Kojibiose	ND
Laminaribiose	3.2
Turanose	5.5
Raffinose	3.3
Lactose	8.1
Lactulose	ND
Leucrose	ND
Theanderose	ND

[a] Yields are expressed as the percentage incorporation of the acceptor supplied. Dextran, isomaltose, leucrose and free glucose were also produced in all cases in varying amounts. ND, not determined.

Source: Reprinted from Rastall, R.A. and Bucke, C., *Biotechnology and Genetic Engineering Reviews*, Intercept, Andover, 1992. With permission.

Salmonella, Escherichia coli and *Vibrio cholera* attacking the gut epithelia. It has been found that mannans can block this interaction by themselves binding to the bacterial lectins, and oligosaccharides known to be rich in mannans derived from yeast cell walls have been proposed as additives to chicken, pig and rabbit feed. It has been claimed that lower levels of pathogen and better performance result from this. It is also known that mannans can enhance the immune response which may also help to prevent infection.

Barley diets benefit by the use of β-glucanases in chicken feeds, apparently by an indirect effect on the viscosity of the digesta, which

TABLE 6.3   Acceptor and Non-Acceptor Molecules for the Cyclodextrin Glucosyl Transferases of *Bacillus* sp.

Acceptors	Non-Acceptors
D-Glucose	D-Fructose
L-Sorbose	D-Mannose
D-Xylose	D-Glucosamine
D-Galactose	N-Acetyl-D-glucosamine
2-Deoxy-D-glucose	L-Rhamnose
3-O-Methyl-D-glucose	D-Allose
6-Deoxy-D-glucose	D-Quinovose
Methyl-α-glucoside	D-Glucuronic acid
Methyl-β-glucoside	L-Arabinose
Phenyl-α-glucose	D-Ribose
Phenyl-β-glucoside	α-D-Glucose-1-phosphate
Maltose	D-Glucose-6-phosphate
Nigerose	Trehalose
Sophorose	2-Deoxy-D-*arabinohexose*
Sucrose	*myo*-Inositol
Maltulose	Glucitol
Palatinose	Xylitol
Cellobiose	Glycerol
p-Nitrophenylmaltoside	
6^2-α-Maltosylmaltose	
6^3-α-Glucosylmaltotriose	

Source: Reprinted from Rastall, R.A. and Bucke, C., *Biotechnology and Genetic Engineering Reviews,* 1992. With permission.

leads to improved absorption of virtually all the nutrients. It is also possible that digestion provides extra absorbable sugars, but this is less significant. Similar effects can be seen with pigs, but the main use of this approach is to reduce the differences that may otherwise be found between different batches of barley, since these vary considerably in β-glucan content.

Pentosanases have also been tried. Pentosans have a general reputation as antinutritional factors, and hydrolysis does improve digestibility, though exactly why is not clear. Xylanases and glucanases also have effects in wheat diets, and given the probability that all the activities are present to some extent in commercial enzyme preparations it is not easy to say precisely which enzyme is the main one. The effect is most likely to be indirect via changes in digesta viscosity.

Attempts to use cellulases have been limited by the poor performance of available enzymes in breaking down cell wall structures, while amylases have not found much application in the feed industry. This may be because the animals have no difficulty in degrading starch in any case. For the time being, the main commercial enzyme use is confined to β-glucanases and pentosanases in chick feed in the United States.

6.10 POLYSACCHARIDE AND GLYCOCONJUGATE VACCINES

Certain types of pathogenic bacteria such as strains of *Streptococcus pneumoniae*, *Neisseria meningitidis* (Meningococcus) and *Haemophilus influenzae* type B 'HiB' besides producing harmful or dangerous toxins also produce high-molecular-weight capsular polysaccharides, which themselves appear harmless (Harding et al., 2012). They do though help the bacterium establish an infection and help hide cell surface components from immune recognition and complement activation. Purified extracts of polysaccharides may themselves be immunogenic and can be used at least in principle to produce immunity against the organism that is producing them. As a result, vaccines against these strains from these organisms are now available. The polysaccharides themselves consist of repeat sequences of saccharide residues. One residue that appears frequently is *N*-acetyl neuraminic acid (Figure 6.11), a type of 'sialic acid'. This residue is common on the surface of membrane glycoproteins and also on mucosal surfaces, so considerable care has to be taken against possible autoimmunity problems (a condition where the body's immune system attacks its own cells). Another potentially serious problem with polysaccharide vaccines is that their effects are not generally long lasting: the repeat sequences produce a T-cell-independent short-term IgM rather than long-term IgG response in infants with little immunological memory effect and are ineffective for infants.

The solution has been to couple the polysaccharide chain(s) to a special type of carrier protein to help stimulate a T-cell (memory) response (Astronomo and Burton, 2010). An example is the tetanus toxoid (toxoid = deactivated form of toxin), which has an ellipsoid conformation of aspect ratio ~3:1 (Abdelhameed et al., 2012). The mechanism by which a glyco-conjugate is constructed around this has been described in Nature's *Scientific Reports* journal (Abdelhameed et al., 2016) – that study also showed for HiB glycoconjugates it is the polysaccharide which dictates its physical properties in solution (Figure 6.12).

6.10.1 STABILITY

Critical to the effectiveness of these constructs is their stability with time and storage. Not only do they have to be effective at the moment they are produced, but they have to be effective by the time they are administered to a patient: in between many weeks may have passed, during which they will have been stored warm, cold or frozen. The shock of freeze–thawing

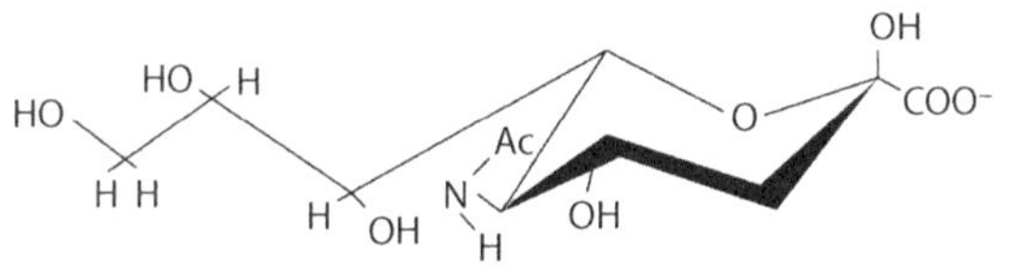

FIGURE 6.11   *N*-acetyl neuraminic acid.

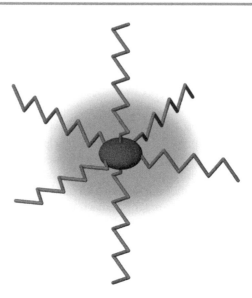

FIGURE 6.12 Schematic diagram of a glycoconjugate-vaccine construct of capsular polysaccharide chains connected to a globular protein core. The flexible polysaccharide chains control the physical properties in vaccine formulations and appear to provide a protective cloud (shade) against protein denaturation and maintaining vaccine effectiveness. (From Abdelhameed, A.S., et al., *Nature Scientific Reports*, 6, 22208, 2016. Courtesy of John Harding.)

itself poses great stresses on the constructs due to the formation of ice crystals and the huge concentration changes that take place. This problem not only applied to gycoconjugate vaccines but many polysaccharide formulations used in biotechnology, particularly in biopharma. In this regard, many of the techniques described in Chapter 1 are invaluable in establishing the molecular integrity of such substances. The outstanding progress in the biotechnology of polysaccharides over the last few decades owes as much to the development of innovative techniques as to the insights that follow their application.

FURTHER READING

Xanthan

Kang, K.S. and Pettit, D.J. (1993) Xanthan, gellan, welan and rhamsan, in Whistler, R.L. and BeMiller, J. N. (eds) *Industrial Gums*, Chap. 13, New York, NY: Academic Press.
Lapasin, R. and Pricl, S. (1995) *Rheology of Industrial Polysaccharides. Theory and Applications*, Chaps. 1 and 2, London: Blackie.

Cyclodextrins

Bender, H. (1986) Production, characterisation and application of cyclodextrins, *Advances in Biotechnological Processes*, 6, 31–71.

Polysaccharide and Glycoconjugate Vaccines

Lepenies, H. ed. (2015) Carbohydrate-Based Vaccines: Methods and Protocols. *Methods in Molecular Biology* volume 1331, Humana Press, New York.

SPECIFIC REFERENCES

Abdelhameed, A.S., Adams, G.G., Morris, G.A., Almutairi, F.M., Duvivier, P., Conrath, K. and Harding, S.E. (2016) A glycoconjugate of Haemophilus influenzae type b capsular polysaccharide with tetanus toxoid protein: Hydrodynamic properties mainly influenced by the carbohydrate, *Nature Scientific Reports*, 6, 22208.

Abdelhameed, A.S., Morris, G.A., Adams, G.G., Rowe, A.J., Laloux, O., Cerny, L., Bonnier, B., et al. (2012) An asymmetric and slightly dimerized structure for the tetanus toxoid protein used in glycoconjugate vaccines, *Carbohydrate Polymers*, 90, 1831–1835.

Astronomo, R.D. and Burton, D.R. (2010) Carbohydrate vaccines: Developing sweet solutions to sticky situations? *Nature Reviews Drug Discovery*, 9, 308–324.

Biver, C., Lesec, J., Allain, C., Salome, L. and Lecourtier, J. (1986) Rheological behaviour and low-temperature sol-gel transition of scleroglucan solutions, *Polymer Communications*, 27, 351–353.

Dea, I.C.M., Morris, E.R., Rees, D.A., Welsh, E.J., Barnes, H.A. and Price, J. (1977) Associations of like and unlike polysaccharides: Mechanism and specificity in galactomannans interacting with bacterial polysaccharides and related systems, *Carbohydrate Research*, 57, 249–272.

Dhami, R., Harding, S.E., Jones, T., Hughes, T., Mitchell, J.R. and To, K. (1995) Physico-chemical studies on a commercial food-grade xanthan. I. Characterisation by sedimentation velocity, sedimentation equilibrium and viscometry, *Carbohydrate Polymer*, 27, 93–99.

Ertesvag, H., Valla, S. and Skjåk-Bræk, G. (1996) Genetics and biosynthesis of alginates, *Carbohydrates in Europe*, 14, 14–18.

Griffin, A.M., Edwards, K.J., Gasson, M.J. and Morris, V.J. (1996) Identification of structural genes involved in bacterial exopolysaccharide production. *Biotechnology and Genetic Engineering Reviews*, 13, 1–18.

Grollier, J.F. (1984) Patent Application (UK) GB 2,136,389A.

Harding, S.E., Abdelhameed, A.S., Morris, G.A., Adams, G.G., Laloux, O., Cerny, L., Bonnier, B., Duvivier, P., Conrath, K. and Lenfant, C. (2012) Solution properties of capsular polysaccharides from *Streptococcus pneumoniae*, *Carbohydrate Polymers*, 90, 237–242.

Hay, I.D., Rehman, Z.U., Moradali, M.F., Wang, Y. and Rehm, B.H.A. (2013) Microbial alginate production, modification and its applications, *Microbial Biotechnology*, 6, doi:10.1111/1751-7915.12076.

Kawahara, K., Ohta, K., Miyamoto, H. and Nakamura, S. (1984) Preparation and solution properties of pullulan fractions as standard samples for water-soluble polymers, *Carbohydrate Polymers*, 4, 335–356.

Kitamura, S., Hori, T., Kurita, K., Takeo, K., Hara, C., Itoh, W., Tabata, K., Elgsaeter, A. and Stokke, B.T. (1994) An antitumor, branched (1→3)-beta-D-glucan from a water extract of fruiting bodies of *Cryptoporus volvatus*, *Carbohydrate Research*, 263, 111–121.

Kitamura, S., Takeo, K., Kuge, T. and Stokke, B.T. (1991) Thermally induced conformational transitions of double-stranded xanthan in aqueous salt solutions, *Biopolymers*, 31, 1243–1255.

Kurachi, K., Ohno, N. and Yadomae, T. (1990) Preparation and antitumor-activity of hydroxyethylated derivatives of 6-branched (1→3)-β-D-glucan obtained from the culture filtrate of *Sclerotina sclerotiorum ifo-9395*, *Chemical and Pharmaceutical Bulletin*, 38, 2527–2531.

Massengale, R., Quinn, F.J., Williams, A. and Aronoff, S.C. (2000) The effect of alginate on the invasion of cystic fibrosis respiratory epithelial cells by clinicalk isolates of *Pseudomonas aeruginosa, Experimental Lung Research*, 26, 163–178.

May, T.B. and Chakrabarty, A.M. (1994) *Pseudomonas aeruginosae*: Genes and enzymes of alginate synthesis, *Trends in Microbiology*, 2, 151–157.

Melia, C.D. (1991) Hydrophilic matrix sustained release systems based on polysaccharide carriers, *Critical Reviews Therapeutic Drug Carrier System*, 8, 395–421.

Millane, R.P., Narasaiah, T.V. and Arnott, S. (1989) On the molecular structures of xanthan and genetically engineered xanthan variants with truncated sidechains, in Crescenzi, V., Dea, I.C.M., Paoletti, S., Stivala, S.S. and Sutherland, I.W. (eds) *Biomedical and Biotechnological Advances in Industrial Polysaccharides*, pp. 469–478, New York, NY: Gordon & Breach.

Morris, V.J. (1987) New and modified polysaccharides, in King, R.D. and Cheetham, P.S. (eds) *Food Biotechnology*, Vol. 1, Chap. 5, London: Elsevier.

Morris, G.A. and Harding, S.E. (2009) Polysaccharides, Microbial, in Schaechter, M., (ed) *Encyclopedia of Microbiology*, 3rd Edition, pp. 482–494, Amsterdam, the Netherlands: Elsevier.

Ogston, A.G. and Silpanata, J. (1970) Thermodynamics of interaction between sephadex and penetrating solute, *Biochemical Journal*, 116, 171–175.

Pindar, D.E. and Bucke, C. (1975) The biosynthesis of alginic acid by *Azotobacter vinelandii, Biochemical Journal*, 152, 617–622.

Rastall, R.A. and Bucke, C. (1992) Enzymatic synthesis of oligosaccharides, in Tombs, M.P. (ed) *Biotechnology and Genetic Engineering Reviews*, Vol. 10, pp. 253–282, Andover: Intercept.

Sanderson, G.R. (1982) The interactions of xanthan gum in food systems, *Progress in Food and Nutrition Science*, 6, 77–87.

Skov-Sørensen, U.B., Blom, J., Birch-Andersen, A. and Henrichsen, J. (1988) Ultrastructural localizatuin of capsules, cell wall polysaccharide, cell wall proteins and F antigen in pneumococci, *Infection and Immunity*, 56, 1890–1896.

Stokke, B.T., Elgsæter, A. and Kitamura, S. (1993) Macrocyclization of polysaccharides visualized by electron microscopy, *International Journal of Biological Macromolecules*, 15, 63–68.

Stokke, B.T., Smidsrød, O. and Elgsæter, A. (1989) Electron microscopy of native xanthan and xanthan exposed to low ionic strength, *Biopolymers*, 28, 617–637.

Sutherland, I.W. (1989) Microbial polysaccharides—biotechnological products of current and future potential, in Crescenzi, V., Dea, I.C.M., Paoletti, S., Stivala, S.S. and Sutherland, I.W. (eds) *Biomedical and Biotechnological Advances in Industrial Polysaccharides*, pp. 123–132, New York, NY: Gordon & Breach.

Vanderslice, R.W., Doherty, D.H., Capage, M.A., Betlach, M.R., Hassler, R.A., Henderson, N.M., Ryan-Graniero, J. and Tecklenburg, M. (1989) Genetic engineering of polysaccharide structure in *Xanthomonas campestris*, in Crescenzi, V., Dea, I.C.M., Paoletti, S., Stivala, S.S. and Sutherland, I.W. (eds) *Biomedical and Biotechnological Advances in Industrial Polysaccharides*, pp. 145–156, New York, NY: Gordon & Breach.

Yanaki, T., Norisuye, T. and Fujita, H. (1980) Triple helix of *Schizophyllum* commune polysaccharide in dilute solution 3. Hydrodynamic properties in water, *Macromolecules*, 13, 1462–1466.

Index